AF780825

Numerical Modeling in Energy and Environment

Numerical Modeling in Energy and Environment

Editors

María Isabel Lamas Galdo
Hassane Naji
Juan de Dios Rodríguez García

MDPI • Basel • Beijing • Wuhan • Barcelona • Belgrade • Manchester • Tokyo • Cluj • Tianjin

Editors
María Isabel Lamas Galdo
Universidade da Coruña
Spain

Hassane Naji
University of Artois
France

Juan de Dios Rodríguez García
Universidade da Coruña
Spain

Editorial Office
MDPI
St. Alban-Anlage 66
4052 Basel, Switzerland

This is a reprint of articles from the Special Issue published online in the open access journal *Applied Sciences* (ISSN 2076-3417) (available at: https://www.mdpi.com/journal/applsci/special_issues/Numerical_Modeling_Energy_Environment).

For citation purposes, cite each article independently as indicated on the article page online and as indicated below:

LastName, A.A.; LastName, B.B.; LastName, C.C. Article Title. *Journal Name* **Year**, *Volume Number*, Page Range.

ISBN 978-3-0365-6279-7 (Hbk)
ISBN 978-3-0365-6280-3 (PDF)

© 2023 by the authors. Articles in this book are Open Access and distributed under the Creative Commons Attribution (CC BY) license, which allows users to download, copy and build upon published articles, as long as the author and publisher are properly credited, which ensures maximum dissemination and a wider impact of our publications.
The book as a whole is distributed by MDPI under the terms and conditions of the Creative Commons license CC BY-NC-ND.

Contents

About the Editors

María Isabel Lamas Galdo

María Isabel Lamas Galdo has a Ph.D. in Industrial Engineering. Since 2008 she has worked as an Associate Professor at the Higher Polytechnic University College of University of A Coruña, Spain. Moreover, she has collaborated with Norplan Engineering S.L. Her main research interest concerns computational fluid dynamics (CFD), specifically applied to thermal systems, marine engines, and pollution reduction. She has written several books, has authored dozens of publications in ISI journals and has participated in several national and international conferences and research projects. She is also experienced in engineering projects.

Hassane Naji

Hassane Naji is a full Professor of Mechanical Engineering at Artois University (France). He holds his Ph.D. in Physical Sciences from the University of Lille (France), and an M.Sc. in Thermal-Fluid Sciences at the National Polytechnic Institute of Lorraine (INPL), Nancy, France. He was a guest researcher (1 year) at the Reacting Gas Dynamics Laboratory, Massachusetts Institute of Technology (MIT), Cambridge, MA 02139, USA. His research covers computational engineering with application to thermal-fluid sciences, advanced heat transfers and numerical methods for multi-scale simulations (mesoscopic and macroscopic approaches).

He is a fellow of various professional institutes and societies (e.g., the American Society of Thermal and Fluids Engineers (ASTFE), the Thermal French Society, etc.) and is regularly invited as a keynote speaker at national and international conferences. He has supervised over 100 M.Sc., Ph.D. and postdoctoral students, many of whom are leaders in academia, industry and even government; published several referenced articles (>100) and conferences (>300) in leading journals and conferences, and lectured extensively worldwide. He is a referee for many leading scientific journals.

Juan de Dios Rodríguez García

Juan de Dios Rodríguez García has a Ph.D. in Industrial Engineering. Since 2001 he has been working as an Associate Professor at the Ferrol Polytechnic School of Engineering (University of A Coruña, Spain). He previously worked as an engineer in engineering and electrical assembly and instrumentation companies. His main research interest refers to computational fluid dynamics (CFD), specifically applied to thermal systems and energy simulation in buildings. He is the author of publications in ISI journals and has participated in various conferences and national and international research projects.

applied sciences

MDPI

Editorial

Numerical Modeling in Energy and Environment

María Isabel Lamas Galdo

Escola Politécnica de Enxeñaría de Ferrol, Universidade da Coruña, C/Mendizábal s/n, 15403 Ferrol, Spain; isabel.lamas.galdo@udc.es

Nowadays, numerical methods constitute an important tool in the analysis of information that cannot be obtained experimentally, or that can be obtained only at a high cost or subject to significant disadvantages. The evolution that computation has undergone in recent years has made it possible to analyze cases that could not be possible some years ago.

Numerical models can be applied to a huge variety of fields, among which one particularly interesting application is in the area of energy and environment. In light of this, the Special Issue "Numerical Models in Energy and Environment" was proposed to collect contributions related to this topic such as two- and three-dimensional modeling in general, computational fluid dynamics, finite element analyses, mathematical models, advanced models, new application areas, etc. A total of 10 manuscripts were published in this Special Issue, with contributions from Spain, Italy, Canada, India, Poland, Portugal, Belgium, China, Germany, Ukraine and Czech Republic.

Several works were aimed at renewable energies, such as Damota et al. [1,2], who analyzed a vertical axis wind turbine Savonius type with CFD (computational fluid dynamics). Nguyen et al. [3] developed two models to improve the precision of estimating the operating temperature of outdoor photovoltaic modules. Shen et al. [4] conducted experimental tests and numerical calculations to analyze the hydraulic vibration and possible exciting sources of analysis in a hydropower system. Other papers were focused on thermal aspects, such as the work of Lebre et al. [5], who developed a computational modeling of the thermal behavior of a greenhouse, and Simon et al. [6], who focused on radiative transfer in complex environments. Other interesting contributions were those of Fomin et al. [7], who determined the vertical load on the carrying structure of a flat wagon with the 18–100 and Y25 bogies; Santos et al. [8], who evaluated the thermal performance and energy efficiency of CRAC equipment through mathematical modeling using a new index COP WEUED; Pandey et al. [9], who carried out CFD investigations of cyclone separators with different cone heights and shapes; and finally Badiozamani and Beier [10], who estimated the potential differential settlement of a tailing deposit based on consolidation properties heterogeneity.

Conflicts of Interest: The author declare no conflict of interest.

Citation: Galdo, M.I.L. Numerical Modeling in Energy and Environment. *Appl. Sci.* **2023**, *13*, 24. https://doi.org/10.3390/app13010024

Received: 9 December 2022
Accepted: 14 December 2022
Published: 20 December 2022

Copyright: © 2022 by the author. Licensee MDPI, Basel, Switzerland. This article is an open access article distributed under the terms and conditions of the Creative Commons Attribution (CC BY) license (https://creativecommons.org/licenses/by/4.0/).

References

1. Blanco Damota, J.; Rodríguez García, J.D.D.; Couce Casanova, A.; Telmo Miranda, J.; Caccia, C.G.; Galdo, M.I.L. Optimization of a Nature-Inspired Shape for a Vertical Axis Wind Turbine through a Numerical Model and an Artificial Neural Network. *Appl. Sci.* **2022**, *12*, 8037. [CrossRef]
2. Blanco, J.; Rodriguez, J.; Couce, A.; Lamas, M. Proposal of a Nature-Inspired Shape for a Vertical Axis Wind Turbine and Comparison of Its Performance with a Semicircular Blade Profile. *Appl. Sci.* **2021**, *11*, 6198. [CrossRef]
3. Nguyen, D.P.N.; Neyts, K.; Lauwaert, J. Proposed Models to Improve Predicting the Operating Temperature of Different Photovoltaic Module Technologies under Various Climatic Conditions. *Appl. Sci.* **2021**, *11*, 7064. [CrossRef]
4. Shen, A.; Chen, Y.; Zhou, J.; Yang, F.; Sun, H.; Cai, F. Hydraulic Vibration and Possible Exciting Sources Analysis in a Hydropower System. *Appl. Sci.* **2021**, *11*, 5529. [CrossRef]
5. Lebre, B.; Silva, P.D.; Pires, L.C.; Gaspar, P.D. Computational Modeling of the Thermal Behavior of a Greenhouse. *Appl. Sci.* **2021**, *11*, 11816. [CrossRef]
6. Simon, H.; Sinsel, T.; Bruse, M. Advances in Simulating Radiative Transfer in Complex Environments. *Appl. Sci.* **2021**, *11*, 5449. [CrossRef]
7. Fomin, O.; Lovska, A.; Píštěk, V.; Kučera, P. Determination of the Vertical Load on the Carrying Structure of a Flat Wagon with the 18–100 and Y25 Bogies. *Appl. Sci.* **2021**, *11*, 4130. [CrossRef]
8. Santos, A.; Gaspar, P.; de Souza, H. Evaluation of the Thermal Performance and Energy Efficiency of CRAC Equipment through Mathematical Modeling Using a New Index COP WEUED. *Appl. Sci.* **2021**, *11*, 5950. [CrossRef]
9. Pandey, S.; Saha, I.; Prakash, O.; Mukherjee, T.; Iqbal, J.; Roy, A.K.; Wasilewski, M.; Brar, L.B. CFD Investigations of Cyclone Separators with Different Cone Heights and Shapes. *Appl. Sci.* **2022**, *12*, 4904. [CrossRef]
10. Badiozamani, M.M.; Beier, N. Estimating the Potential Differential Settlement of a Tailings Deposit Based on Consolidation Properties Heterogeneity. *Appl. Sci.* **2022**, *12*, 6206. [CrossRef]

Disclaimer/Publisher's Note: The statements, opinions and data contained in all publications are solely those of the individual author(s) and contributor(s) and not of MDPI and/or the editor(s). MDPI and/or the editor(s) disclaim responsibility for any injury to people or property resulting from any ideas, methods, instructions or products referred to in the content.

applied sciences

MDPI

Article

Optimization of a Nature-Inspired Shape for a Vertical Axis Wind Turbine through a Numerical Model and an Artificial Neural Network

Javier Blanco Damota [1], Juan de Dios Rodríguez García [1], Antonio Couce Casanova [1], Javier Telmo Miranda [2], Claudio Giovanni Caccia [3] and María Isabel Lamas Galdo [1,*]

1 Escuela Politécnica de Ingeniería de Ferrol, Universidade da Coruña, 15403 Ferrol, Spain
2 Escuela Técnica Superior de Ingenieros Industriales, Universidad Nacional de Educación a Distancia (UNED), 28040 Madrid, Spain
3 Department of Aerospace Engineering, Politecnico di Milano, 20156 Milan, Italy
* Correspondence: isabel.lamas.galdo@udc.es; Tel.: +34-881013896

Abstract: The present work proposes an artificial neural network (ANN) to analyze vertical axis wind turbines of the Savonius type. These turbines are appropriate for low wind velocities due to their low starting torque. Nevertheless, their efficiency is too low. In order to improve the efficiency, several modifications are analyzed. First of all, an innovative blade profile biologically inspired is proposed. After that, the influence of several parameters such as the aspect ratio, overlap, and twist angle was analyzed through a CFD (computational fluid dynamics) model. In order to characterize the most appropriate combination of aspect ratio, overlap, and twist angle, an artificial neural network is proposed. A data set containing 125 data points was obtained through CFD. This data set was used to develop the artificial neural network. Once established, the artificial neural network was employed to analyze 793,881 combinations of different aspect ratios, overlaps, and twist angles. It was found that the maximum power coefficient, 0.3263, corresponds to aspect ratio 7.5, overlap/chord length ratio 0.1125, and twist angle 112°. This corresponds to a 32.4% increment in comparison to the original case analyzed with aspect ratio 1, overlap 0, and twist angle 0.

Keywords: wind turbines; VAWT; CFD; Savonius; Fibonacci; ANN

Citation: Blanco Damota, J.; Rodríguez García, J.d.D.; Couce Casanova, A.; Telmo Miranda, J.; Caccia, C.G.; Lamas Galdo, M.I. Optimization of a Nature-Inspired Shape for a Vertical Axis Wind Turbine through a Numerical Model and an Artificial Neural Network. *Appl. Sci.* **2022**, *12*, 8037. https://doi.org/10.3390/app12168037

Academic Editor: Guan Heng Yeoh

Received: 23 July 2022
Accepted: 8 August 2022
Published: 11 August 2022

Publisher's Note: MDPI stays neutral with regard to jurisdictional claims in published maps and institutional affiliations.

Copyright: © 2022 by the authors. Licensee MDPI, Basel, Switzerland. This article is an open access article distributed under the terms and conditions of the Creative Commons Attribution (CC BY) license (https://creativecommons.org/licenses/by/4.0/).

1. Introduction

Nowadays, energy demand is constantly increasing. Finite resources such as fossil fuels are being depleted, and global pollution is reaching alarming levels. These facts lead to the current necessity to urgently implement renewable resources. In this regard, wind energy plays a crucial role in the decarbonization process of society [1–4] and constitutes the most relevant energy contributor to the renewable sector [5]. Depending on the orientation of the rotation axis, wind turbines can be categorized into HAWTs (horizontal axis wind turbines) and VAWTs (vertical axis wind turbines). HAWTs are the most employed ones, mainly installed in high-power wind farms. An important disadvantage is that HAWTs are too dependent on wind conditions [6–8]. On the other hand, VAWTs are less common but also interesting due to their capacity to capture wind from all directions and their acceptable performance under poor wind conditions. These characteristics make VAWTs appropriate for small power generation applications, especially in urban environments [9]. VAWTs are basically categorized into Darrieus (based on lift) and Savonius (based on drag). The Savonius VAWTs are recommended when the wind velocity is low due to their lower starting torque, i.e., they need lower wind velocities to start [10]. Nevertheless, the efficiency of the Savonius turbine is lower than the Darrieus turbine. In order to increment the efficiency of the Savonius turbine, several parameters have been analyzed in the literature. Several augmentation techniques have been proposed, such as nozzles, deflectors, guide vanes, end plates, etc. Another improvement of the Savonius turbine relies on the blade profile. Some authors proposed modifications to the semicircular blade profile [11–16]. The efficiency can also be increased through some parameters related to the rotor design, such as the aspect ratio (AR), overlap (O), separation gap (SG), twist angle (TA), number of blades, etc. [17–19]. Figure 1a illustrates a 3D view of a Savonius turbine. It is composed of semicircular blades vertically positioned along the rotation axis. The O is the distance that an inside edge penetrates into another. The SG is the width separating the inside edges of the blades. The TA is the angle between the upper and lower sections of the blades. The turbine shown in Figure 1a has a zero value of AR, O, SG, and TA. The AR, O, and SG are illustrated in Figure 1b and the TA in Figure 1c. The AR is the relation between the height and the diameter, AR = H/D.

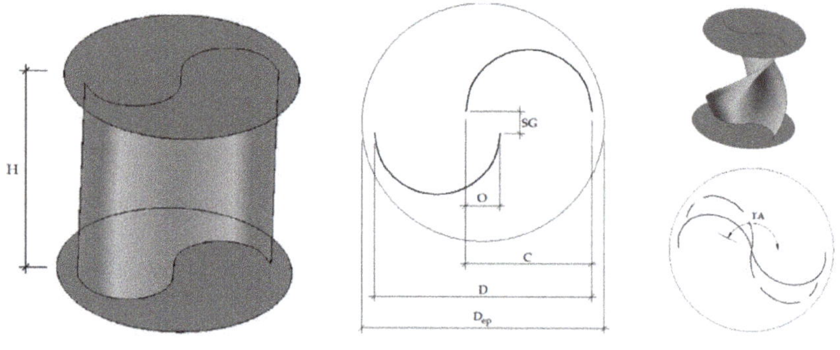

(a) (b) (c)

Figure 1. Savonius turbine; (**a**) 3D view; (**b**) lateral section of a Savonius turbine with non-zero overlap and separation gap; (**c**) illustration of the twist angle. H: height, C: chord length, O: overlap, SG: separation gap, D: rotor diameter, D_{ep}: end plate diameter, TA: twist angle.

Unfortunately, the relations between these parameters and the efficiency are too complex, and there is no analytical relation between them. The determination of the most appropriate combination of AR, SG, O, etc., is a complex nonlinear problem and there is a huge number of combinations. Because of this, a tool is required to efficiently predict the power output. In this regard, artificial neural networks (ANNs) may constitute a useful solution for these types of problems. In fact, ANNs are recommended for problems that are too complex to solve and where enough data can be obtained. The principle of operation of ANNs consists of simulating the biological neuron structure and learning capacity. ANNs employ artificial neurons similarly to biological neurons in the human brain. Through the training process, an ANN takes the input/s variable/s and inter-relates them to obtain the output/s variable/s. An important advantage is that a well-trained ANN is too fast to predict the output variable/s. ANNs are able to learn the relationships between the inputs and outputs and provide output predictions with a higher accurate prediction than traditional methods [20,21]. The nonlinear application of ANNs enables them to address many problems that are very difficult to be modeled mathematically. ANNs avoid complex physical and/or mathematical models. In recent years, ANNs have been widely used in many fields, such as engineering, science, economy, etc.

After analyzing several models and taking into account the tendency observed in the literature, the present work proposes a nature-inspired blade profile to be used in VAWTs, particularly a blade profile inspired by Fibonacci's spiral, based in turn on Fibonacci's mathematical sequence. This is presented in many nature scenes, such as the formation of some flowers and fruits, hurricanes, and even the galaxies with their rotation movements. In a previous work [22], the Fibonacci shape was proposed to improve the efficiency of the conventional Savonius turbine. Subsequently, the effect of the number of blades, aspect ratio, overlap, separation gap, and twist angle was illustrated [23], but a method to obtain the most appropriate combination of these parameters was not proposed. In order to solve this issue, the preset work proposes to employ an ANN to obtain the most appropriate combination of parameters. In total, 125 training data points were used to develop the ANN, and this dataset was obtained through CFD simulations. The ANN predicts the power coefficient and was employed to analyze 793,881 cases. The maximum power coefficient of these alternatives was extracted and proposed as the most appropriate option.

The main objectives of the present work are:

- Proposing an innovative blade profile for a Savonius-type VAWT based on the Fibonacci spiral.
- Carrying out a CFD model validated with experimental measurements in order to provide 125 data to establish an ANN.
- Establishing an ANN to analyze 793,881 possible combinations of aspect ratio, overlap, and twist angle and determine the most appropriate combination of all these cases considered.

The remainder of this paper is structured as follows. Section 2 describes the methodology employed in the numerical model and the ANNs. Section 3 presents the results obtained and a discussion about them. Finally, the conclusions of this work are indicated in Section 4.

2. Materials and Methods

This section describes the CFD model employed to obtain the data used to train the ANN and its corresponding validation using experimental results. Moreover, the advantages of the new blade profile are illustrated. After that, the methodology used to establish the ANN is described.

2.1. CFD Analysis

The CFD procedure carried out to simulate the VAWT was described in a previous paper [23]. For this reason, it will be briefly described herein. The model is based on the RANS equations of conservation of mass and momentum. The turbulence was treated through the SST k-ω turbulence model. The software OpenFOAM was employed. An important step in the CFD analysis was the validation process. To this end, data obtained by Sandia laboratories [24] were employed. These experiments are based on the turbine shown in Figure 1a. The main characteristics are listed in Table 1. This turbine has two blades, 1 m diameter, 1 m height, 0 overlap, 0 separation gap, and 0 twist angle. Two end plates were added to reduce the escape of air from the blades.

Table 1. Characteristics of the turbine employed for the validation procedure.

Parameter	Value
Diameter (m)	1
Height (m)	1
Number of blades	2
Overlap (m)	0
Separation gap (m)	0
Twist angle (°)	0

The mesh employed is shown in Figure 2. Two zones are differenced: static and rotating. The static domain is almost the whole region, while the rotating domain is a cylinder that rotates according to the rotational velocity. The turbine is placed inside the rotating domain. Regarding boundary conditions, a velocity inlet was modeled on the upstream face, while the downstream face was modeled as an outlet. The exterior faces were assumed to free sleep and the turbine surface was modeled as no-slip.

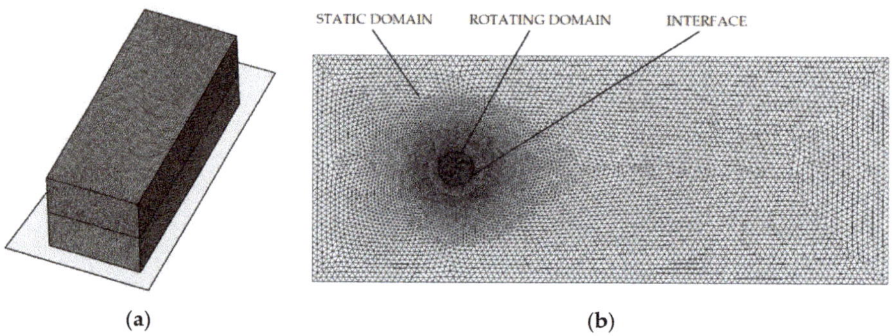

Figure 2. Computational mesh. (**a**) 3D view; (**b**) middle plane.

A comparison between the experimental and numerical results is shown in Figure 3. Particularly, Figure 3a shows the power coefficient (C_P) and Figure 3b compares the torque coefficient (C_T) against the tip speed ratio (TSR). These parameters represent the non-dimensioned form of the rotational velocity, power, and torque, and are given by Equations (1)–(3), respectively. In these equations, V is the free-stream velocity, ω the rotational speed, P the power, ρ the density of air, ν the kinematic viscosity of air, S the cross-section area (S = DH), and T the torque. Figure 3 shows an adequate concordance between experimental and numerical results. Several reasons are responsible for the differences between experimental and numerical data. First of all, CFD is an approximate method and unavoidable errors are introduced due to the discretization processes and hypotheses realized to simplify the model. On the other hand, experimental measurements are affected by the tolerance of the apparatus employed. Moreover, another important source of error is the blockage effect. This effect is produced when a sample is placed in the test section of a wind tunnel; the sample alters the current velocity. Several methods can be found in the literature to correct this modified current velocity, but significant differences can be obtained using different methods [25].

$$TSR = \frac{\text{blade tip tangential velocity}}{\text{wind speed}} = \frac{\omega R}{V} \tag{1}$$

$$C_P = \frac{\text{power}}{\text{available power}} = \frac{P}{0.5\rho S V^3} \tag{2}$$

$$C_T = \frac{\text{torque}}{\text{available torque}} = \frac{T}{0.25\rho S V^2 D} \tag{3}$$

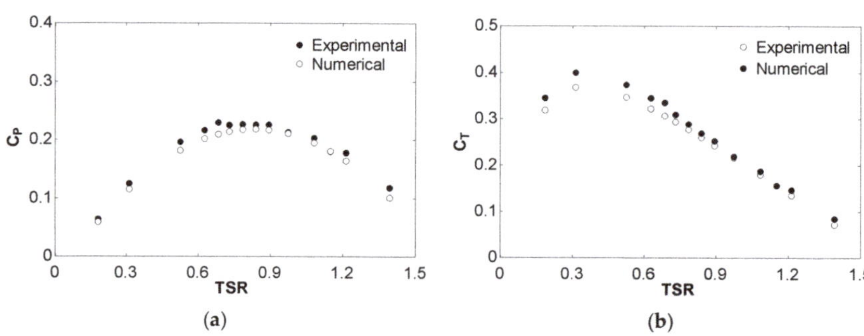

Figure 3. Comparison between numerical and experimental results; (**a**) C_P against TSR; (**b**) C_T against TSR.

Once the model was validated, the first improvement proposed in the present work is to modify the semicircular blade profile originally proposed by Sigur Savonius [26]. The nature-inspired Fibonacci spiral is proposed to this end. This shape is highly presented in nature in animals, plants, and several natural phenomena—Figure 4. Since this shape is the result of thousands of centuries of evolution, it is reasonable to propose this geometry for VAWT Savonius-type turbines. It is worth mentioning that some of the authors of the present work have previously proposed bioinspired shapes to be applied in engineering mechanisms with successful results [27–30].

Figure 4. Fibonacci spiral in a seashell.

The Fibonacci spiral is composed of quarter circumferences. The higher the consecutive terms, the closer the relationship is to the so-called divine proportion or golden ratio, i.e., the terms R_{n-1} and R_n shown in Figure 5a tend to the relation $R_{n-1}/R_n = 1.61803398874896$ … if R_{n-1} and R_n are high enough. The resulting turbine is shown in Figure 5b. This ratio between two consecutive terms comes from dividing a segment into two parts, a and b, where a is longer than b, and the relation between the total length to a is the relation between a and b, as shown in Figure 5c. The resulting succession is 0, 1, 1, 2, 3, 5, 8, 13, 21, 34, 55, 89, etc. The first terms of this succession do not fulfill the divine proportion, but they are closer to the divine proportion as higher are the terms.

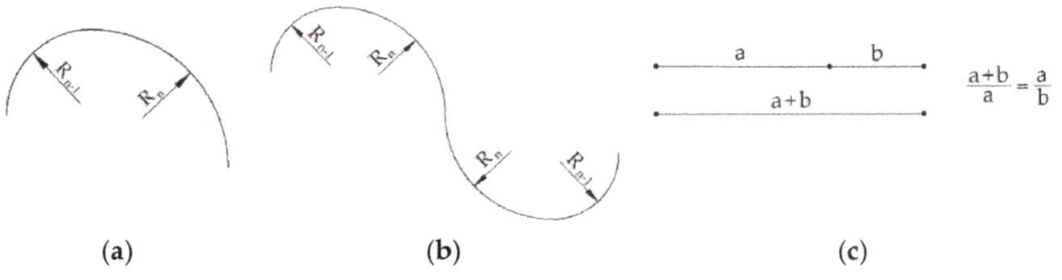

Figure 5. Fibonacci spiral; (**a**) two consecutive terms; (**b**) turbine obtained from Fibonacci spirals used in the present work; (**c**) graphical representation of the golden ratio.

According to Figure 5c, and taking into account that a must be a positive number, the length of a is obtained through the following development:

$$\frac{a+b}{a} = \frac{a}{b} \rightarrow a^2 = ab + b^2 \rightarrow a = \frac{b + \sqrt{b^2 + 4b}}{2} \tag{4}$$

And the relation between a and b, i.e., the divine proportion or golden ratio is:

$$\frac{a}{b} = \frac{1 + \sqrt{1+4}}{2} = 1.61803398874896\ldots \tag{5}$$

Figure 6a,b show a comparison between the Fibonacci and original Savonius blade profiles. These profiles are illustrated in the figures, using a grey color dotted line for the original Savonius and a black color dashed line for the Fibonacci. Particularly, Figure 6a illustrates the power coefficient, and Figure 6b compares the torque coefficient against the TSR. It can be seen that the proposed blade profile improves the original Savonius since it provides higher power and torque coefficients.

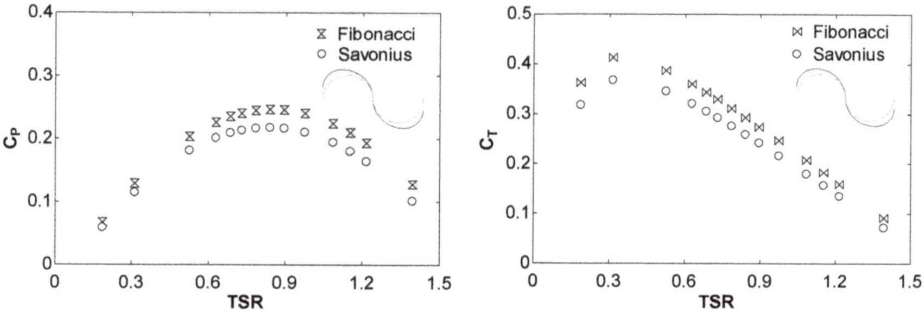

(a) (b)

Figure 6. Comparison between the Fibonacci and original Savonius blade profiles; (**a**) C_P against TSR; (**b**) C_T against TSR.

It is worth mentioning that the values shown in Figure 6 correspond to the average value along a rotor cycle (360°). The variation of the power and torque coefficients along a cycle is shown in Figure 7, which illustrates these values for a TSR = 0.84. Other values of TSR provide similar results and thus are not shown again.

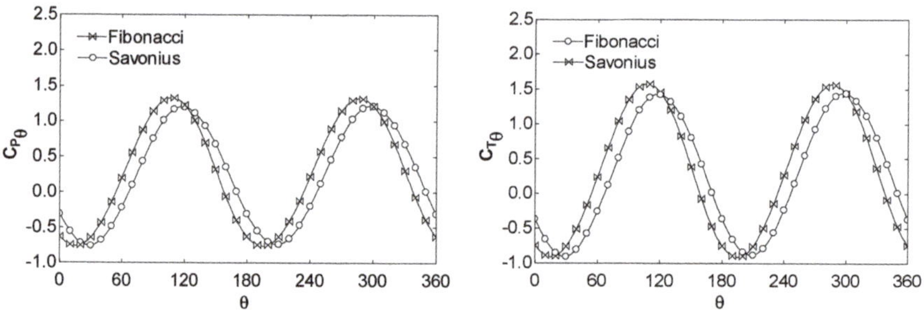

(a) (b)

Figure 7. Performance of Fibonacci and original Savonius blade profiles along a rotor cycle; (**a**) $C_{P\theta}$; (**b**) $C_{T\theta}$.

2.2. *ANN Analysis*

Once the superiority of the Fibonacci blade profile was verified through CFD, the next step is to analyze the effect of other parameters. In a previous work [22], it was verified that the optimum number of blades is 2 and the optimum separation gap is 0. The same conclusion was obtained by other authors [17,31–33]. Nevertheless, the optimum aspect ratio, overlap, and twist angle are not clear, and there is no consensus in the scientific literature. Further on, the determination of the most appropriate combination of these parameters is a complex issue. The reason is that a huge number of combinations are possible and the relation between these parameters and the power is not a simple task. As mentioned above, in order to analyze a high number of combinations, an ANN is proposed in the present work.

Three input variables were established: the aspect ratio, overlap, and twist angle. It is worth mentioning that it is common to provide the overlap as the overlap/chord length ratio, i.e., O/C. Regarding the output variables, in this case, only one variable is needed, the maximum power coefficient, C_{Pmax}. For instance, regarding the base case with AR = 1,

O/C = 0, and TA = 0, the maximum power coefficient is the maximum value of the graph shown in Figure 6a, which corresponds to 0.2465. According to this, the ANN used for the present work will perform as a "black box" in which the three inputs are provided and the output is returned, as shown in Figure 8.

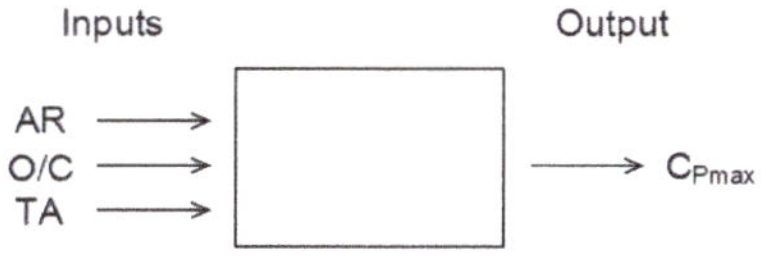

Figure 8. ANN structure for the present work.

ANNs are structures that imitate human intuition. An ANN is capable of simulating the human brain and is able to process information and provide the corresponding predictions. The human brain uses neurons to process data. Figure 9 shows the structure of biological neurons. The data are transferred from synapses to the axon through electrochemical media called neurotransmitters. The human brain contains approximately 100 billion interconnected neurons [34]. The information in the brain propagates via electrochemical media known as neurotransmitters and is transferred along the neurons. The artificial neural network imitates this configuration. The principle of operation of ANNs is based on interconnected nodes that can process information. Two main advantages are provided by ANNs:

- Capacity to learn from existing data and thus the ability to provide a prediction from multivariable relationships between process parameters.
- Capacity to treat complex relationships between dependent and independent variables with high accuracy.

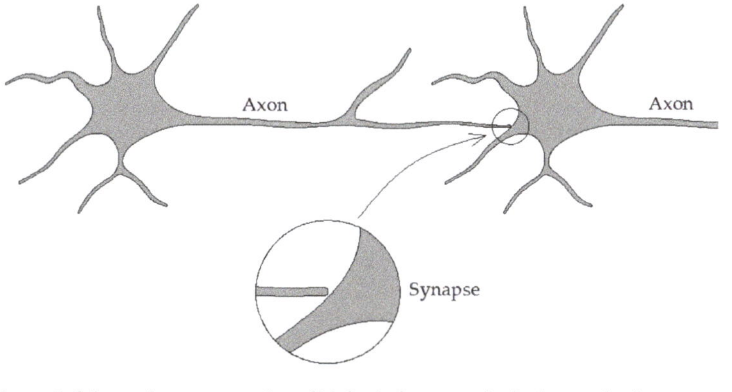

Figure 9. Schematic representation of biological neurons in the human brain.

In the present work, the software Matlab 2021 was employed. The learning between the inputs and outputs of an ANN is performed through a mathematical training process that minimizes errors and provides an optimal prediction. The neurons of the ANN are organized into three layers: input, hidden, and output—Figure 10. The input layer corresponds to the independent variables; in this particular work, the aspect ratio, overlap/chord length ratio, and twist angle. On the other hand, the output layers correspond to the dependent variable; in this particular work, C_{Pmax}. The effect of the input to the cell refers to weights. There is no exact rule to define the number of hidden layers and hidden nodes, and several methods to determine them can be found in the literature. Generally, a single hidden layer is recommended for most problems, and the multi-layered structure is only recommended for complex problems [35–38] since adding hidden layers may cause memorizing instead of generalizing [39]. Each input node has an assigned weight and transfer functions related to the nodes.

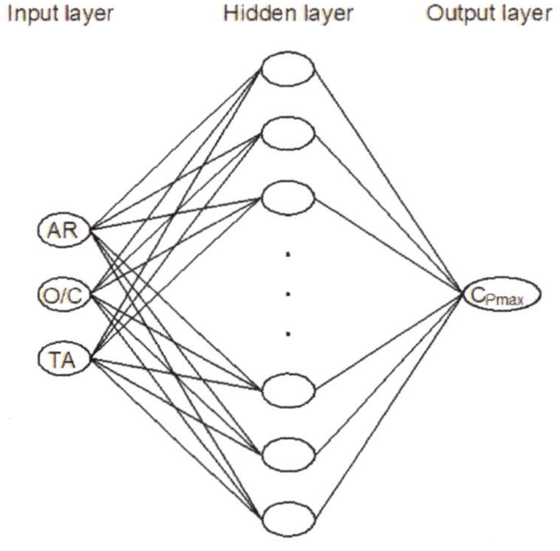

Figure 10. Structure of the ANN employed for the present analysis.

Two common problems in ANNs are underfitting and overfitting. An ANN must accurately fit the input and output data, as shown in Figure 11a. Underfitting takes place when the ANN is too simple, or the data sample is too small. In this case, the ANN does not accurately fit the data, as shown in Figure 11b. On the other hand, overfitting takes place when the model memorizes instead of generalizing. In this case, the ANN fits too well during the training process but poorly on the test dataset [40], as shown in Figure 11c. Regarding the number of neurons in the hidden layer, a low quantity may lead to high computational cost and overfitting, while a low number of neurons may lead to underfitting. In the present work, several networks with hidden layer neurons between 3 and 20 were tested.

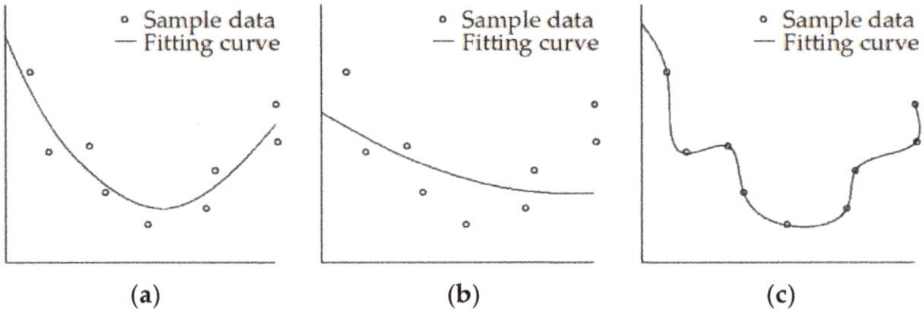

Figure 11. Data fitting performance; (**a**) appropriate fitting; (**b**) underfitting; (**c**) overfitting.

3. Results

In order to establish the ANN, a data set must be provided. As mentioned previously, these data set were obtained through the CFD model. A total of 125 cases were analyzed using different values of the aspect ratio, overlap, and twist angle. Particularly, five values of the aspect ratio were employed: 1, 3, 5, 7, and 9; five values of the overlap/chord length ratio: 0, 0.05, 0.1, 0.15, and 0.2; and five values of the twist angle: 0°, 60°, 120°, 180°, and 240°. These ranges were chosen according to proposed values in the literature. Taking into account these values of the aspect ratio, overlap/chord length ratio, and twist angle, a total of $5 \times 5 \times 5 = 125$ cases were analyzed. These cases are graphically represented in Figure 12.

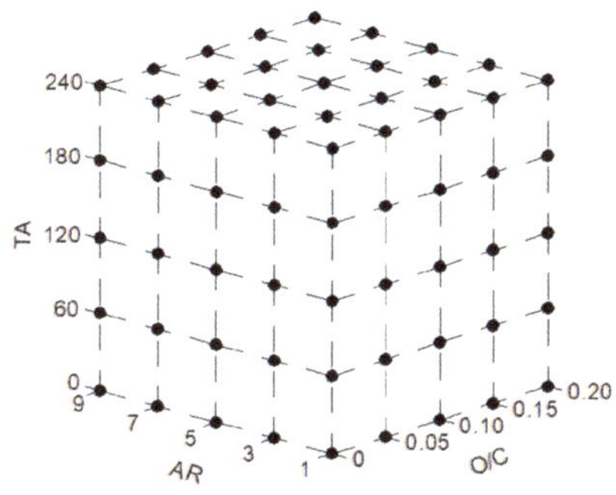

Figure 12. Graphical representation of the 125 cases analyzed through the CFD model.

The CFD provided the power coefficients of all these 125 cases. Some results of the maximum power coefficients are illustrated in Figures 13–15. Particularly, Figure 13 shows the maximum power coefficient against O/C and AR, Figure 14 shows the maximum power coefficient against TA and O/C, and Figure 15 shows the maximum power coefficient against AR and TA. It can be seen that the influence of the aspect ratio, overlap, and twist angle on the power coefficient is not a simple task since some values of the aspect ratio increment the maximum power coefficient but, from a certain value, the maximum power coefficient decreases again. The same effect is observed regarding the overlap and twist angle. These relations make it difficult to achieve the most appropriate combination of the aspect ratio, overlap, and twist angle. The only way to solve it appropriately is to simulate a huge quantity of possible combinations. Nevertheless, this procedure would take too much time with the computational resources currently available. This fact justifies the employment of alternative methods such as ANNs.

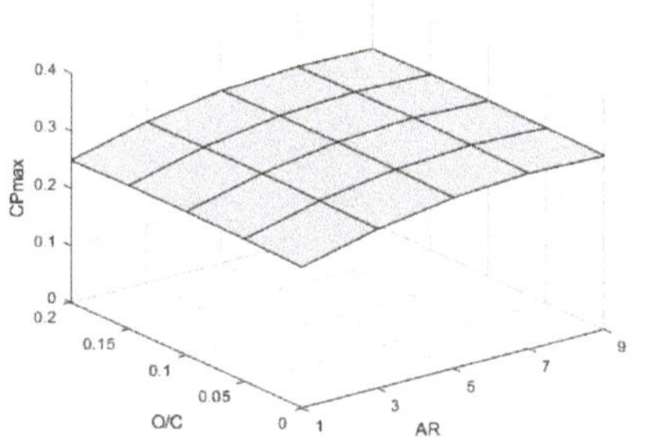

Figure 13. Maximum power coefficient against O/C and AR. TA = 0.

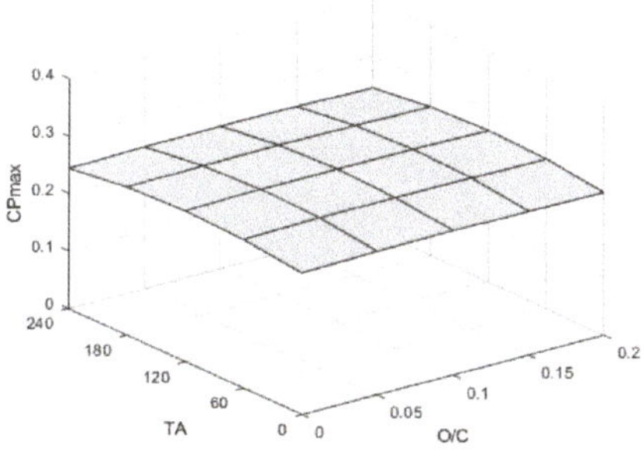

Figure 14. Maximum power coefficient against TA and O/C. AR = 1.

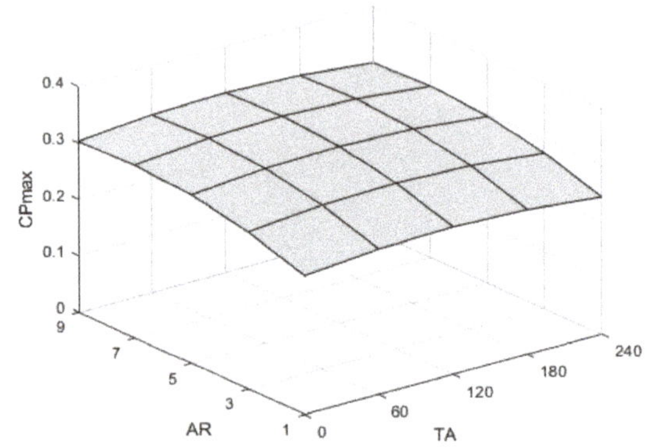

Figure 15. Maximum power coefficient against AR and TA. O/C = 0.

Once these 125 data were obtained through CFD, they will be employed to set the ANN and analyze a considerably higher amount of data. In order to minimize the domination of any parameter, data standardization was realized. This standardization or normalization process eliminates the dimensions and scales the data in a range, usually from 0 to 1, Equation (6).

$$x_n = \frac{x - x_{min}}{x_{max} - x_{min}} \quad (6)$$

where x_n is the normalized value, x_{max} is the maximum x value, and x_{min} is the minimum x value.

Once normalized, these 125 cases analyzed through CFD were employed as samples to train, validate, and test the network. Particularly, 89 samples, randomly selected, were used to train the network, 18 to validate it, and 18 to test it. Figure 14 shows the performance separated into three sets: training, validation, and test. Moreover, all data are also provided. The accuracy is evaluated on the basis of the prediction errors from the verification data. Equation (7) is used to compute the correlation coefficient R. The global R-value is too close to the optimum value R = 1. The closer to R = 1, the more accurate the ANN is. Graphically, Figure 16 shows that the fitting of the four images is practically diagonal, thus providing an excellent data fitting. It can be seen that the curves of the four images are basically on the diagonal, which indicates an appropriate data fitting, i.e., an accurate prediction of the maximum power coefficient.

$$R = \sqrt{1 - \frac{\sum_{i=1}^{m} \left[(t_i - o_i)^2 \right]}{\sum_{i=1}^{m} o_i^2}} \quad (7)$$

where t is the target value, o the output, and m the number of samples.

As mentioned previously, this ANN was employed to analyze a much higher quantity of cases than 125. Particularly, 81 values of the aspect ratio were employed: 1, 1.1, 1.2, ... 9; 81 values of the overlap/chord length ratio: 0, 0.0025, 0.005, ... 0.2; and 121 values of the twist angle: 0°, 2°, 4°, ... 240°. Taking into account these values of the aspect ratio, overlap/chord length, and twist angle, a total of 81 × 81 × 121 = 793,881 cases were analyzed. The most appropriate solution obtained by the ANN is shown in Table 2. It can be seen that the optimum solution corresponds to AR = 7.5, O/C = 0.1125, and TA = 112°, providing a maximum power coefficient of 0.3263. Regarding the aspect ratio, high values increment the power coefficient, but excessive values reduce the positive effect that the endplates promote on the performance (reducing escaping of air). Regarding the overlap, it is positive since it allows the flowing of air from the advancing to the returning blade, incrementing the power. Nevertheless, excessive overlaps lead to a high reduction in the air on the advancing blade, which reduces the power. Regarding the twist angle, the optimum value also improves the aerodynamics of the turbine. For comparison, the base case with AR = 1, O/C = 0, and TA = 0 is also shown in Table 2. The obtained solution improves the maximum power coefficient by 32.4% in comparison with the base case.

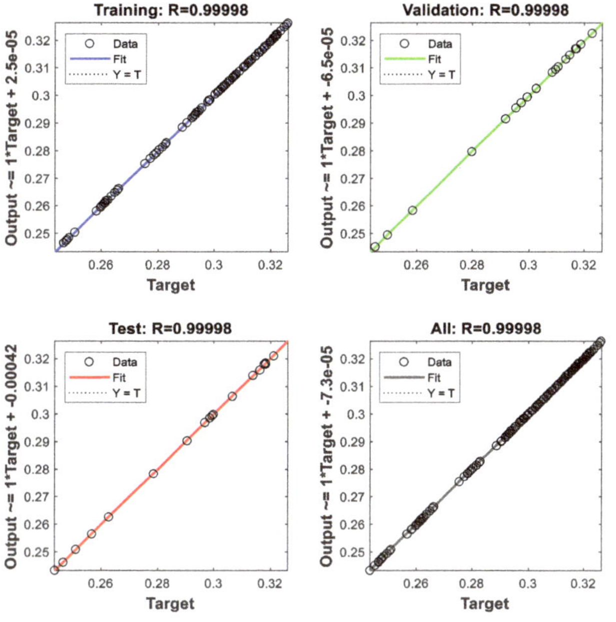

Figure 16. Regression graphs of the ANN.

Table 2. Optimum and base case solutions.

	AR	O/C	TA	C_{pmax}
Base case	1	0	0	0.2465
Optimum	7.5	0.1125	112	0.3263

For comparison purposes, Figure 17 compares the power coefficient against the TSR corresponding to the base and optimum cases. It can be seen that the improvement is considerable.

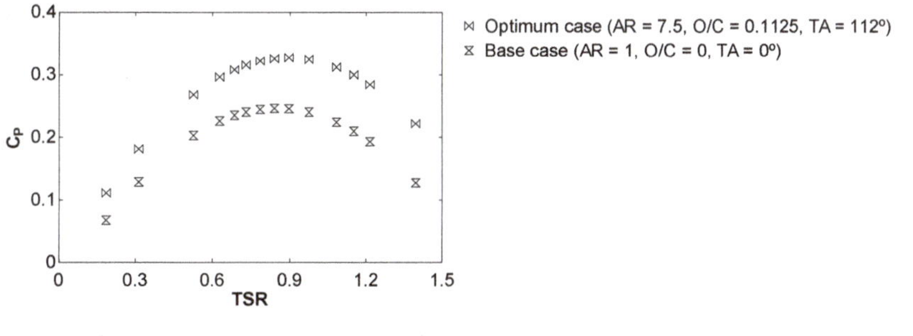

Figure 17. Comparison between the base and the optimum cases.

4. Conclusions

The present work focuses on an ANN to analyze a Savonius-type VAWT. The reason is the current necessity to employ renewable resources to reduce both global pollution and the dependency on finite resources such as fossil fuels. The Savonius turbine is appropriate for small power generations, but its main disadvantage relays on its low efficiency. With a proper design, it is possible to increment its efficiency. According to this, the present work proposes a bio-inspired blade profile based on the Fibonacci spiral. Previous studies showed that the performance of these turbines could be affected by several parameters. The present work focuses on the aspect ratio, overlap, and twist angle that influence the efficiency of the Savonius turbine. The variety of possible combinations of these parameters is immense; moreover, their relation to efficiency and thus optimization is a complex nonlinear problem. To predict the efficiency, an ANN was developed. The data employed to set the ANN were obtained through a validated CFD model. One hundred twenty-five cases were employed to train, validate, and test the neural network. The ANN was validated using R-value, which was close to 1. This shows that the ANN can accurately predict the performance as an alternative to classical modeling techniques and direct measurements. Once established, the ANN was used to analyze a total of 793,881 cases using different combinations of aspect ratios, overlap/chord length ratios, and twist angles. It was found that the most appropriate combination corresponds to aspect ratio 7.5, overlap/chord length ratio 0.1125, and twist angle 112°. This combination improves the maximum power coefficient by 32.4% in comparison to the base case with aspect ratio 1, overlap/chord length ratio 0, and twist angle 0. Regarding future works, it will be interesting to analyze other improvements such as deflectors, several stages, and combinations of Savonius–Darrieus stages.

Author Contributions: Conceptualization, J.d.D.R.G., J.B.D., J.T.M., A.C.C., M.I.L.G. and C.G.C.; methodology, J.d.D.R.G., J.B.D., J.T.M., A.C.C., M.I.L.G. and C.G.C.; software, J.d.D.R.G., J.B.D., J.T.M., A.C.C., M.I.L.G. and C.G.C.; validation, J.d.D.R.G., J.B.D., J.T.M., A.C.C., M.I.L.G. and C.G.C.; formal analysis, J.d.D.R.G., J.B.D., J.T.M., A.C.C., M.I.L.G. and C.G.C.; investigation, J.d.D.R.G., J.B.D., J.T.M., A.C.C., M.I.L.G. and C.G.C.; resources, J.d.D.R.G., J.B.D., J.T.M., A.C.C., M.I.L.G. and C.G.C.; writing—original draft preparation, J.d.D.R.G., J.B.D., J.T.M., A.C.C., M.I.L.G. and C.G.C.; writing—review and editing, J.d.D.R.G., J.B.D., J.T.M., A.C.C., M.I.L.G. and C.G.C.; supervision, J.d.D.R.G., J.B.D., J.T.M., A.C.C., M.I.L.G. and C.G.C. All authors have read and agreed to the published version of the manuscript.

Funding: This research received no external funding.

Institutional Review Board Statement: Not applicable.

Informed Consent Statement: Not applicable.

Conflicts of Interest: The authors declare no conflict of interest.

References

1. Hegler, S.; Plettemeier, D. Simulative investigation of the radar cross section of wind turbines. *Appl. Sci.* **2019**, *9*, 4024. [CrossRef]
2. Ding, L.; Guo, T. Numerical study on the power efficiency and flow characteristics of a new type of wind energy collection Device. *Appl. Sci.* **2020**, *10*, 7438. [CrossRef]
3. Ancuti, M.C.; Musuroi, S.; Sorandaru, C.; Dordescu, M.; Erdodi, G.M. Wind turbines optimal operation at time variable wind speeds. *Appl. Sci.* **2020**, *10*, 4232. [CrossRef]
4. Dolinski, L.; Krawczuk, M. Analysis of modal parameters using a statistical approach for condition monitoring of the wind turbine blade. *Appl. Sci.* **2020**, *10*, 5878. [CrossRef]
5. Jessen, K.; Laugesen, K.; Mortensen, S.M.; Jensen, J.K.; Soltani, M.N. Experimental validation of aero-hydro-servo-elastic models of a scaled floating offshore wind turbine. *Appl. Sci.* **2019**, *9*, 1244. [CrossRef]
6. Qi, L.; Zheng, L.; Bai, X.; Chen, Q.; Chen, J.; Chen, Y. Nonlinear maximum power point tracking control method for wind turbines considering dynamics. *Appl. Sci.* **2020**, *10*, 811. [CrossRef]
7. Astolfi, D.; Castellani, F.; Berno, F.; Terzi, L. Numerical and experimental methods for the assessment of wind turbine control upgrades. *Appl. Sci.* **2018**, *8*, 2639. [CrossRef]
8. Han, X.; Liu, D.; Xu, C.; Shen, W.; Li, L.; Xue, F. Monin–Obukhov similarity theory for modeling of wind turbine wakes under atmospheric stable conditions: Breakdown and modifications. *Appl. Sci.* **2019**, *9*, 4256. [CrossRef]

9. Wijayanto, R.P.; Kono, T.; Kiwata, T. Performance characteristics of an orthopter-type vertical axis wind turbine in shear flows. *Appl. Sci.* **2020**, *10*, 1778. [CrossRef]
10. Damota, J.; Lamas, I.; Couce, A.; Rodriguez, J. Vertical axis wind turbines: Current technologies and future trends. *Renew. Energy Power Qual. J.* **2015**, *1*, 530–535. [CrossRef]
11. Alom, N.; Saha, U.K. Influence of blade profiles on Savonius rotor performance: Numerical simulation and experimental validation. *Energy Convers. Manag.* **2019**, *186*, 267–277. [CrossRef]
12. Kacprzak, K.; Liskiewicz, G.; Sobczak, K. Numerical investigation of conventional and modified Savonius wind turbines. *Renew. Energy* **2013**, *60*, 578–585. [CrossRef]
13. Kumar, G.; Ram, V.R.; Kumar, N. Numerical analysis of different blade profile of wind turbine. *Int. J. Appl. Eng. Res.* **2018**, *6*, 375–385.
14. Benesh, A.H. Wind Turbine System Using a SAVONIUS-Type Rotor. U.S. Patent 4,715,776A, 27 February 1996.
15. Mohamed, M.H.; Janiga, G.; Pap, E.; Thévenin, D. Optimal blade shape of a modified Savonius turbine using an obstacle shielding the returning blade. *Energy Convers. Manag.* **2011**, *52*, 236–242. [CrossRef]
16. Chan, C.M.; Bai, H.L.; He, D.Q. Blade shape optimization of the Savonius wind turbine using a genetic algorithm. *Appl. Energy* **2018**, *213*, 148–157. [CrossRef]
17. Akwa, J.V.; Vielmo, H.A.; Petry, A.P. A review on the performance of Savonius wind turbines. *Renew. Sustain. Energy Rev.* **2012**, *16*, 3054–3064. [CrossRef]
18. Mahmoud, N.H.; El-Haroun, A.A.; Wahba, E.; Nasef, M.H. An experimental study on improvement of Savonius rotor performance. *Alex. Eng. J.* **2012**, *51*, 19–25. [CrossRef]
19. Zhang, H.; Li, Z.; Xin, D.; Zhan, J. Improvement of aerodynamic performance of Savonius wind rotor using straight-arc curtain. *Appl. Sci.* **2020**, *10*, 7216. [CrossRef]
20. Bu, L.; Du, G.; Hou, Q. Prediction of the compressive strength of recycled aggregate concrete based on artificial neural network. *Materials* **2021**, *14*, 3921. [CrossRef]
21. Asteris, P.G.; Moropoulou, A.; Skentou, A.D.; Apostolopoulou, M.; Mohebkhah, A.; Cavaleri, L.; Rodrigues, H.; Varum, H. Stochastic vulnerability assessment of masonry structures: Concepts, modeling and restoration aspects. *Appl. Sci.* **2019**, *9*, 243. [CrossRef]
22. Blanco, J.; Rodriguez, J.D.; Couce, A.; Lamas, M.I. Proposal of a nature-inspired shape for a vertical axis wind turbine and comparison of its performance with a semicircular blade profile. *Appl. Sci.* **2021**, *11*, 6198. [CrossRef]
23. Damota, J.B.; García, J.D.D.R.; Casanova, A.C.; Miranda, J.T.; Caccia, C.G.; Galdo, M.I.L. Analysis of a nature-inspired shape for a vertical axis wind turbine. *Appl. Sci.* **2022**, *12*, 7018. [CrossRef]
24. Blackwell, B.F.; Sheldahl, R.E.; Feltz, L.V. *Wind Tunnel Performance Data for Two- and Three-Bucket Savonius Rotors*; Sandia Laboratories: Springfield, VA, USA, 1977.
25. Ross, I.J. Wind Tunnel Blockage Corrections: An Application to Vertical-Axis Wind Turbines. Master's Thesis, University of Dayton, Dayton, OH, USA, 2010.
26. Savonius, S.J. The S-rotor and its applications. *Mech. Eng.* **1931**, *53*, 333–338.
27. Lamas, M.I.; Rodriguez, C.G. Hydrodynamics of biomimetic marine propulsion and trends in computational simulations. *J. Mar. Sci. Eng.* **2020**, *8*, 479. [CrossRef]
28. Lamas Galdo, M.I.; Rodriguez Vidal, C.G.; Rodriguez Garcia, J.D. Optimization of the efficiency of a biomimetic marine propulsor using CFD. *Ing. Investig.* **2014**, *34*, 17–21.
29. Lamas, M.I.; Rodriguez, J.D.; Rodriguez, C.G. CFD analysis of biologically-inspired marine propulsors. *Brodogradnja* **2012**, *63*, 125–133.
30. Lamas, M.; Rodríguez, J.; Rodríguez, C.; González, P. Three-dimensional CFD analysis to study the thrust and efficiency of a biologically-inspired marine propulsor. *Pol. Marit. Res.* **2011**, *18*, 10–16. [CrossRef]
31. Sheldahl, R.E.; Blackwell, B.F.; Feltz, L.V. Wind tunnel performance data for two- and three-bucket Savonius rotors. *J. Energy* **1978**, *2*, 160–164. [CrossRef]
32. Saha, U.K.; Thotla, S.; Maity, D. Optimum design configuration of Savonius rotor through wind tunnel experiments. *J. Wind Eng. Ind. Aerodyn.* **2008**, *96*, 1359–1375. [CrossRef]
33. Zhao, Z.; Zheng, Y.; Xu, X.; Liu, W.; Hu, G. Research on the Improvement of the Performance of Savonius Rotor Based on Numerical Study. In Proceedings of the 2009 International Conference on Sustainable Power Generation and Supply, Nanjing, China, 6–7 April 2009; pp. 1–6. [CrossRef]
34. Bhatt, A.N.; Shrivastava, N. Application of artificial neural network for internal combustion engines: A state of the art review. *Arch. Comput. Methods Eng.* **2022**, *29*, 897–919. [CrossRef]
35. Özcan, E.; Danışan, T.; Yumuşak, R.; Eren, T. An artificial neural network model supported with multi criteria decision making approaches for maintenance planning in hydroelectric power plants. *Eksploat. Niezawodność.-Maint. Reliab.* **2020**, *22*, 400–418. [CrossRef]
36. Golmohammadi, D. Neural network application for fuzzy multi-criteria decision making problems. *Int. J. Prod. Econ.* **2011**, *131*, 490–504. [CrossRef]

37. Lamas Galdo, M.I.; Telmo Miranda, J.; Rebollido Lorenzo, J.M.; Caccia, C.G. Internal modifications to optimize pollution and emissions of internal combustion engines through multiple-criteria decision-making and artificial neural networks. *Int. J. Environ. Res. Public Health* **2021**, *18*, 12823. [CrossRef] [PubMed]
38. Rodriguez, C.G.; Lamas, M.I.; Rodriguez, J.D.; Abbas, A. Analysis of the pre-injection system of a marine diesel engine through multiple-criteria decision-making and artificial neural networks. *Pol. Marit. Res.* **2022**, *28*, 88–96. [CrossRef]
39. Sietsma, J.; Dow, R.J.R. Creating artificial neural networks that generalize. *Neural Netw.* **1991**, *4*, 67–79. [CrossRef]
40. Yuan, Z.; Shi, X.; Jiang, D.; Liang, Y.; Mi, J.; Fan, H. Data-based engine torque and NOx raw emission prediction. *Energies* **2022**, *15*, 4346. [CrossRef]

applied sciences

MDPI

Article

Proposal of a Nature-Inspired Shape for a Vertical Axis Wind Turbine and Comparison of Its Performance with a Semicircular Blade Profile

Javier Blanco [1], Juan de Dios Rodriguez [2], Antonio Couce [2] and Maria Isabel Lamas [1,*]

[1] Nautical Sciences and Marine Engineering Department, Hygher Politechnic University College, University of Coruña, 15403 Ferrol, Spain; javier.blanco.damota@udc.es

[2] Industrial Engineering Department, Politechnic University College, Univertisy of Coruña, 15405 Ferrol, Spain; de.dios.rodriguez@udc.es (J.d.D.R.); antonio.coucec@udc.es (A.C.)

* Correspondence: isabel.lamas.galdo@udc.es; Tel.: +34-881-013-896

Featured Application: This work analyzes the Savonius vertical axis wind turbine. A new geometry inspired the Fibonacci's spiral was analyzed, obtaining a considerable improvement by both CFD and experimental tests.

Abstract: In order to improve the efficiency of the Savonius type vertical axis wind turbine, the present work analyzes an improvement based on an innovative rotor geometry. The rotor blades are inspired on an organic shape mathematically analyzed, the Fibonacci's spiral, presented in many nature systems as well as in art. This rotor was analyzed in a wind tunnel and through a CFD model. The power coefficients at different tip speed ratios (TSR) were characterized and compared for the Savonius turbine and two versions using the Fibonacci's spiral. One of the proposed geometries improves the performance of the Savonius type. Particularly, the optimal configuration lead to an improvement in maximum power coefficient of 14.5% in the numerical model respect to a conventional Savonius turbine and 17.6% in the experimental model.

Keywords: wind turbine; VAWT; CFD; Savonius; Fibonacci

Citation: Blanco, J.; Rodriguez, J.d.D.; Couce, A.; Lamas, M.I. Proposal of a Nature-Inspired Shape for a Vertical Axis Wind Turbine and Comparison of Its Performance with a Semicircular Blade Profile. *Appl. Sci.* **2021**, *11*, 6198. https://doi.org/10.3390/app11136198

Academic Editor: Elsa Caetano

Received: 10 May 2021
Accepted: 2 July 2021
Published: 4 July 2021

Publisher's Note: MDPI stays neutral with regard to jurisdictional claims in published maps and institutional affiliations.

Copyright: © 2021 by the authors. Licensee MDPI, Basel, Switzerland. This article is an open access article distributed under the terms and conditions of the Creative Commons Attribution (CC BY) license (https://creativecommons.org/licenses/by/4.0/).

1. Introduction

Nowadays, the energy framework is experiencing a significant and fast change promoted by the awareness on climate change. In this framework, renewable energy plays an important role. Wind-power generation is one of the fundamental sources of renewable energy [1–4], which has become the major energy contributor to the renewable energy sector [5]. It is estimated that by 2030, wind energy will provide about 20% of the world's electricity demand [6]. The main drawback, though, is that the wind resources are affected by wind volatility and, furthermore, sudden wind speed variations lead to an unstable operation of the wind turbine [7–9].

According to the orientation of the rotation axis, wind turbines are classified into horizontal axis wind turbines (HAWTs) and vertical axis wind turbines (VAWTs). VAWT are usually applied to small-scale, especially urban environments characterized by winds not totally appropriate to this end [10–13]. In addition, the power is generated in the place of consumption, reducing transportation losses [14]. In urban environments, the Savonius turbine constitutes one of the most appropriate candidates due to their adequacy to low wind velocities [13]. The Savonius wind turbine is a VAWT composed of two or three arc-type blades which can generate power even under poor wind conditions [15,16] such as bursts, fluctuating wind, and turbulence. In urban environments the Savonius VAWT offers an appropriate alternative since these turbines can start at low wind speeds and in highly turbulent flows. Other advantages are their simplicity, low cost, and independence on wind direction. In addition, these turbines are more easily maintained due to their small size and the fact that the alternator and gearbox can be placed on the ground. On the other hand, the main disadvantages of the Savonius turbines is the low power coefficient and efficiency in comparison with other VAWTs or HAWTs. In order to increase their efficiency and thus the power generation, these turbines have been studied for four decades. One of the most significant improvements consists on blade profiles that are not semicircular, initially proposed by Sigur Savonius [17]. In recent years, important investigations have been carried out to improve the semicircular blade profile. Most of these studies have proposed elliptical blade profiles as optimal geometries to increment the power produced by the vertical axis Savonius turbines [18–24].

After analyzing several models and taking into account the tendency observed in the literature, the present work proposes a nature-inspired blade profile, particularly a blade profile inspired by the Fibonacci's spiral, based in turn on Fibonacci's mathematical sequence. This is presented in many natural contexts such as the formation of some flowers and fruits, hurricanes, and even the galaxies with their rotation movements. The geometry proposed in the present work was analyzed both experimentally and numerically. The experimental analysis were carried out in a wind tunnel. The numerical analysis were developed by a 3D CFD (Computational Fluid Dynamics) model. Two different Fibonacci inspired geometries were compared with the Savonius turbine. The objective of the present work is to advance in the search for a more efficient blade geometry for a Savonius type VAWT. To this end, a modified blade profile has been implemented. The improvement in performance offered by the proposed geometry has been tested both numerically and experimentally, taking as a reference the performance of the VAWT with a semicircular profile. The numerical analysis was developed using a 3D CFD (Computational Fluid Dynamics) model. The corresponding experimental verifications were carried out in a wind tunnel, maintaining the premise of seeking clarity in the comparison despite having to lose precision in the replica of the numerical model. Two different Fibonacci-inspired geometries were compared with the Savonius turbine. In relation to biological inspirations, it is worth mentioning that two of the authors of the present work investigated an innovative propulsion for aquatic devices through biological inspiration in previous works, obtaining successful results [25–27].

2. Materials and Methods

2.1. Blade Profile

The nature-inspired blade profile analyzed in the present work is constituted by two consecutive sections of the Fibonacci's spiral, obtained through Fibonacci's mathematical sequence. The Fibonacci sequence is recurrent, i.e., all previous terms are needed to compute a specific term. Equations (1)–(3) define the recurrence relation:

$$f_0 = 0 \tag{1}$$

$$f_1 = 1, \text{ if } n = 1 \tag{2}$$

$$f_n = f_{n-2} + f_{n-1}, \text{ if } n > 1 \tag{3}$$

where f is the relative square size and n is the number of squares.

The procedure to establish the graphical representation is based on a set of squares whose length is given by the successive terms of the Fibonacci's sequence. These squares must be placed as shown in Figure 1a, where the numbers represent the relative size of each square side. According to Equations (1)–(3), the relative size of the sequence is 0, 1, 1, 2, 3, 5, 8, 12, 21, etc. As can be seen in Figure 1a, the first square has been colored in red. The second square is placed on the right of the previous one, the third square below the second one, the fourth on the left of the third one, and so on. The next step consists on drawing an arc as shown in Figure 1b.

(a) (b)

Figure 1. Fibonacci's spiral; (**a**) distribution of the squares; (**b**) spiral with rations.

This work proposes a 200 mm height rotor based on two Fibonacci spirals corresponding to the consecutive terms 34 and 55 of the sequence. Two support endplates have been incorporated. These dimensions have been chosen according to the size of the wind tunnel, 400 × 400 mm, used for the experimental setup. The goal was to obtain as much clarity as possible in the comparison. The higher the rotors are, the clearer the contrast is, but the accuracy is reduced due to the blockage effect [28]. The position of both these arcs relative to the rotating axis leads to two different configurations, Figure 2a,b.

(a) (b)

Figure 2. Profiles analyzed; (**a**) Fibonacci I; (**b**) Fibonacci II.

Both profiles shown in Figure 2a,b, Fibonacci I and Fibonacci II, respectively, have been analyzed. The characteristics of these geometries are shown in Table 1. The parameters shown in this table are illustrated in Figure 3. As can be seen, the configuration Fibonacci II is obtained from Fibonacci I by simply inverting the order of the arcs, leading to a lower rotor diameter.

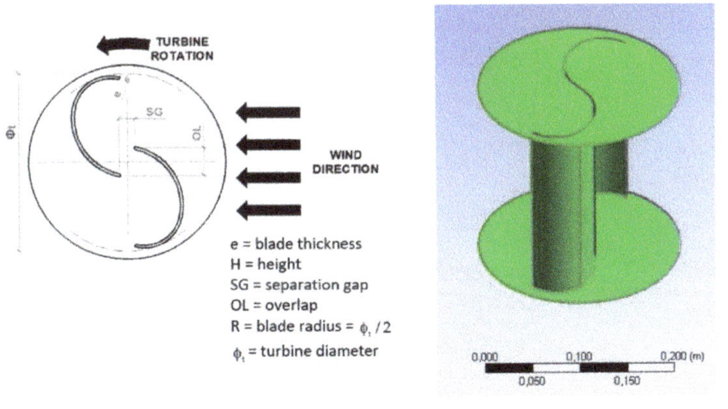

(a) (b)

Figure 3. (**a**) Constructive parameters; (**b**) 3D model.

Table 1. Characteristics of the analyzed wind turbines.

Model	ϕ_t (mm)	R (mm)	H (mm)	$\phi_{endplate}$ (mm)	OL (mm)	SG (mm)	e (mm)
Fibonacci I	194.55	97.27	200	214.0	0	0	3
Fibonacci II	191.66	95.83	200	210.8	0	0	3

2.2. Numerical Model

A CFD model was carried out to analyze the wind turbines. This model is based on the RANS equations (Reynolds-averaged Navier–Stokes) of conservation of mass and momentum. The turbulence was treated through the k-ε turbulence model. The geometry was realized using Siemens NX and the CFD computation using Ansys Fluent.

The SIMPLE algorithm was chosen for the pressure–velocity coupling and a second order upwind scheme was employed to discretize the governing equations. The convergence was determined by 0.001 residuals as convergence criterion. The maximum number of iterations per time step was established as 20. The temporal treatment was resolved through an implicit method with a constant time step corresponding to 1° of rotor rotation. It was verified that this time step size is small enough to provide accurate results. In order to reach a situation which periodically repeats, i.e., a quasi-steady state, it was necessary to study a long enough interval of time. For the cases studied, it was verified that this state is achieved after approximately twenty revolutions. For this reason, all the results carried out in the present work correspond to the 20th revolution.

The computational mesh is shown in Figure 4a. It was realized using the mesh tool in ANSYS and consists of around 5×10^5 tetrahedral elements. Several size meshes were tested in order to verify its adequacy. The mesh size was refined near the turbine, Figure 4b, and a boundary layer was set near the wall, in order to treat the boundary layer flow. Five layers were set, with a layer growth rate of 1.2. The size of the first layer fulfills y^+ values between 30 and 300, depending on the rotation velocity. This range of y^+ values is adequate to use the standard wall functions, employed in the present work to account for turbulent effects near the wall. Two zones were differenced, internal and external. The internal zone is spherical, 1000 mm diameter, and rotates around the axis. The external zone is static. It is a cube with side size 2000 mm. The inner boundaries of the static domain coincide with the outer boundaries of the rotating domain. A sliding mesh technique was adopted to interpolate the values at the boundary between the rotating and static domains. The initialization is based on a rotating reference frame simulation.

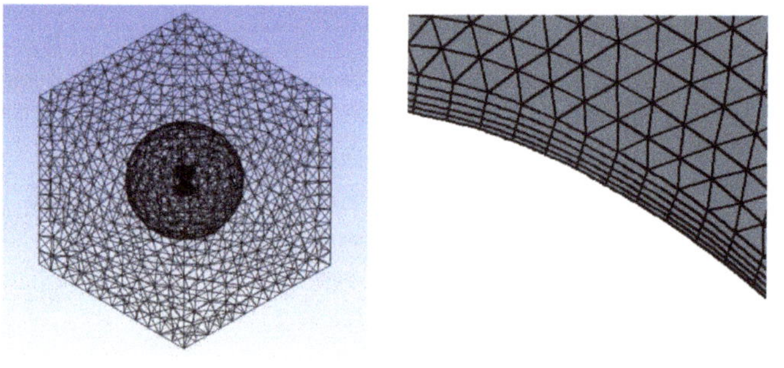

(a) (b)

Figure 4. (a) Computational mesh; (b) boundary layer detail.

The boundary conditions are shown in Figure 5. The upstream surface was modelled as a velocity inlet with 7 m/s velocity, 5% turbulent intensity, and 0.028 m length scale. The downstream surface as pressure outlet, 0 Pa (gauge), 5% turbulent intensity, and 0.028 m length scale. A no-slip condition was imposed at the surface of the blades. An interface was imposed at the overlap surface between the adjacent domains in order to allow the transport of the flow properties. Finally, the external walls were modeled as no-slip.

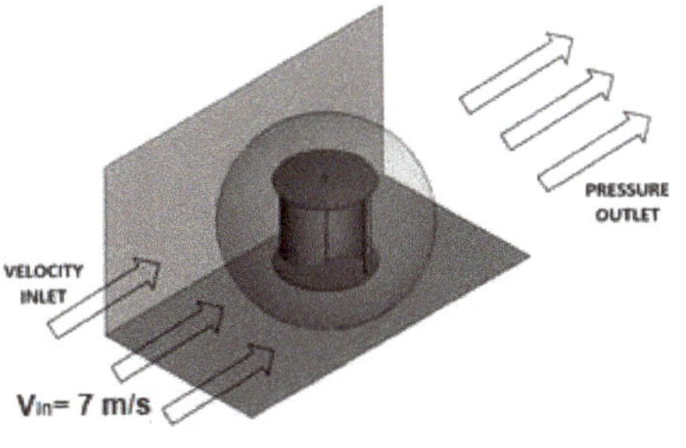

Figure 5. Boundary conditions.

2.3. Experimental Setup

An experimental setup was carried out in the subsonic closed circuit wind tunnel of the Nautical Sciences and Marine Engineering department of University of Coruña, Figure 6a. This tunnel has a 400 × 400 × 800 mm (height × width × length) test section made of transparent glass, Figure 6b.

Figure 6. (**a**) Wind tunnel used for the measurements; (**b**) turbine inside the test section.

A flow fan is controlled through a variable-frequency drive, which provides a maximum speed of 16 m/s. A DC generator was attached to the rotor axis. A 50 Ω and 50 W rheostat was connected after the generator in order to regulate the braking torque in the turbine axis and thus the RPM. The power was characterized through readings of both intensity and voltage by means of a Keysight DSO306A oscilloscope. The readings of wind velocity were obtained through a Pitot tube connected to a KIMO MP200 manometer.

3. Results and Discussion

The present work employs dimensionless parameters to analyze the results. Since the turbines analyzed are scale models of real turbines, a dimensionless analysis allows a comparison with turbines of different sizes and in different wind conditions. TSR is the ratio between of the blade tip tangential velocity and the wind speed, calculated using Equation (4):

$$\text{TSR} = \frac{\text{blade tip tangential velocity}}{\text{wind speed}} = \frac{\omega R}{V} \quad (4)$$

where ω is the rotational speed of the rotor, R is the rotor radius, and V is the wind speed.

C_m is the torque coefficient, defined as follows:

$$C_m = \frac{M}{0.25\rho SV^2\phi_t} \quad (5)$$

where M is the torque, $0.5\rho V^2$ the kinetic energy per unit volume of the incoming flow and S the cross-section area, given by $S = \phi_t H$.

C_p is the power coefficient. It expresses which fraction of the power in the wind is being extracted by the turbine, as shown in Equation (6):

$$C_p = \frac{P}{0.5\rho SV^3} = C_m TSR \quad (6)$$

where P is the shaft power. This should be the power on the shaft, but since the objective of this study is to compare the performance of the Fibonacci rotor against the Savonius rotor, for this case P has been taken as output power (electrical power obtained), taking into account that the losses by friction and electrical are practically the same in both cases

Figure 7 shows the average power coefficients against the TSR for the Fibonacci I, Fibonacci II, and Savonius configurations. Both numerical and experimental results are shown in this figure. Since the Fibonacci II configuration lead to lower TSR than Savonius, it was not considered experimentally. The main difference between Fibonacci I and Fibonacci II is that the former places the lower arc in the end of the blade, Figure 4a, as indicated by the results obtained by other authors, Figure 1.

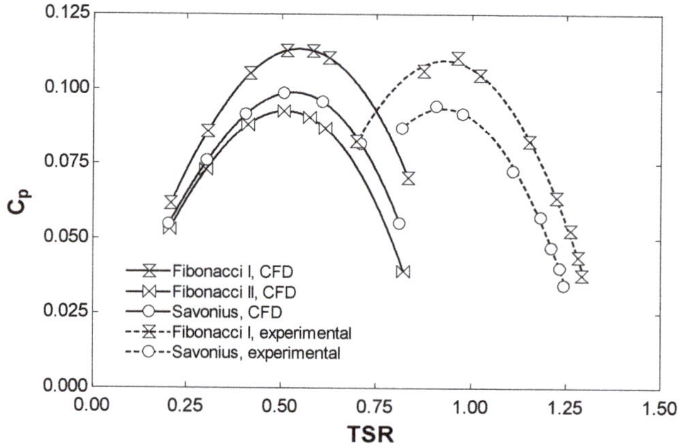

Figure 7. Power coefficients against TSR for Fibonacci I, Fibonacci II, and Savonius configurations.

As can be seen in Figure 7, there is a peak value for each curve and the power coefficient increases with TSR up to a certain point after which it drops down as the TSR further increases. Both the experimental and numerical results show the superiority of the Fibonacci I turbine over the Savonius one. Particularly, the maximum power coefficients for Fibonacci I and Savonius were 0.1129 and 0.0986, respectively, for the CFD analysis, which means a 14.5% improvement of Fibonacci I over Savonius. Regarding the experimental results, the maximum power coefficients for Fibonacci I and Savonius were 0.1106 and 0.0943, respectively, which means a 17.6% improvement of Fibonacci I over Savonius.

Several reasons are responsible for the discrepancies between the numerical and experimental results shown in Figure 7. The mechanical losses in the generator, bearings, transmission, electrical losses in the generator, and the blockage effect do not allow for the experimental reproduction of the CFD data. Regarding CFD, the k-ε turbulence model does not provide accurate results near the wall [29]. DNS (Direct Numerical Simulation) or LES (Large Eddy Simulation) would provide more accurate results but with a considerably higher computational cost. CFD is not an exact science, and the mesh generation and discretization processes induce inevitable numerical errors. Another source of error in the numerical model is the length of the domain. A larger length should provide a better characterization of the wake behind the turbine.

Despite the discrepancies between numerical and experimental results, both procedures demonstrate the superiority of Fibonacci over Savonius, which is the goal of the present work.

Figure 8 shows the power coefficient as a function of the azimuth angle for several TSR values. This figure was obtained through the CFD model. As can be seen, the torque becomes positive as the azimuth angle increases above around 40 and 220° and reaches maximum values around 110 and 290°. The torque becomes negative around 170 and 360° and reaches minimum values around 20 and 200°.

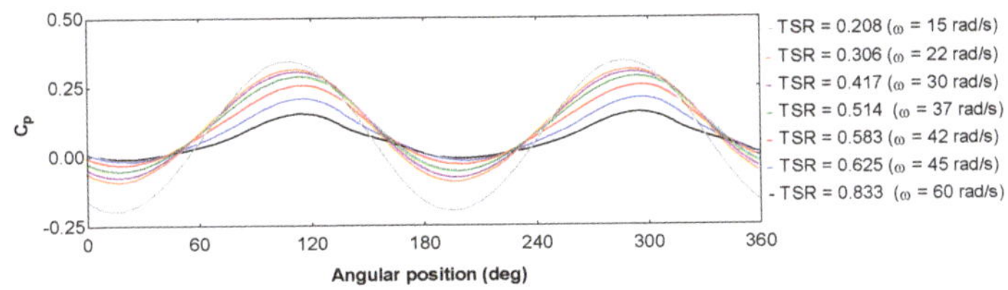

(a)

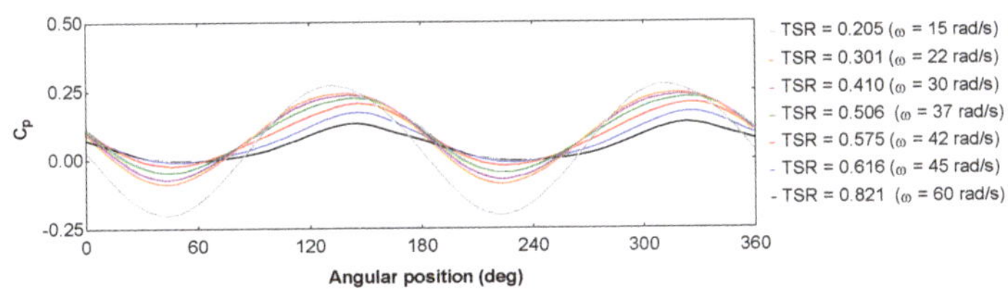

(b)

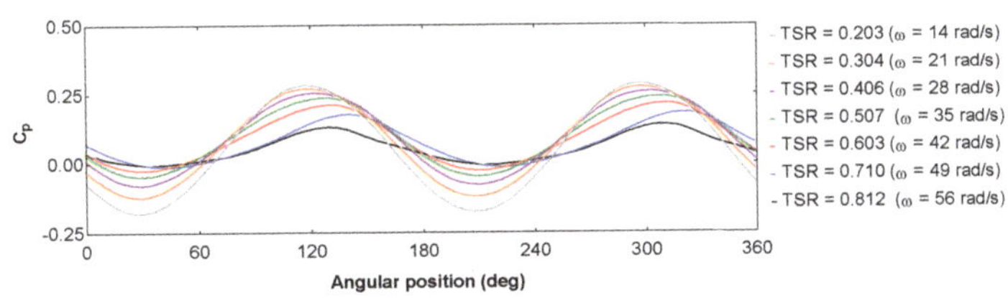

(c)

Figure 8. Power coefficient against the angular position; (**a**) Fibonacci I; (**b**) Fibonacci II; (**c**) Savonius.

Figure 9 shows the velocity and pressure field for the Fibonacci I profile and Figure 10a–c shows the velocity and pressure field in the middle plane for Fibonacci I at TSR = 0.514, Savonius at TSR = 0.507, and Fibonacci II at TSR = 0.506, respectively. As can be seen, a vortex that rotates counterclockwise can be observed. This vortex promotes the pressure gradients that produces power. Another vortex can be seen near the center of the rotor. Since this vortex is placed near the axis, it has little effect on the power generation. As can be seen in Figure 10, the pressure exerted on the recoil blade (rear blade) is lower for the Fibonacci profile, which favors the rotation of the turbine, being able to obtain higher results in the C_p. This effect can be observed in the illustrations facing the turbines with both profiles in the rotation from 0° to 60°, in which it is observed that greater pressure is exerted on the recoil blade for the Savonius profile than for the Fibonacci one, while the opposite occurs in the advance blade or front blade, the pressure exerted on the Savonius profile is less than that exerted on the Fibonacci profile. In turn, the velocity vectors are more uniformly distributed and have a greater incidence in the vicinity of the forward blade in the case of the Fibonacci profile, as well as in the blade peak, whose concentration is higher in the case of the Fibonacci rotor.

Figure 9. Velocity and pressure field for the Fibonacci I profile.

	Fibonacci I	**Savonius**	**Fibonacci II**
0°	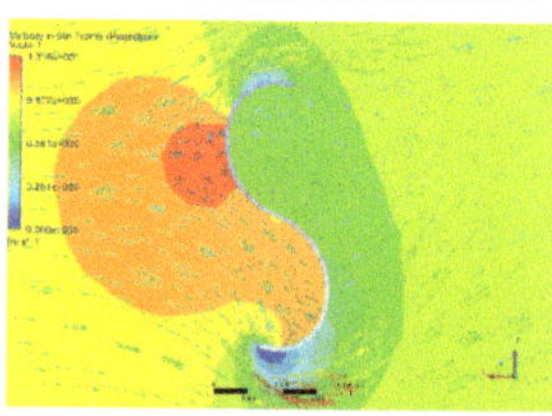	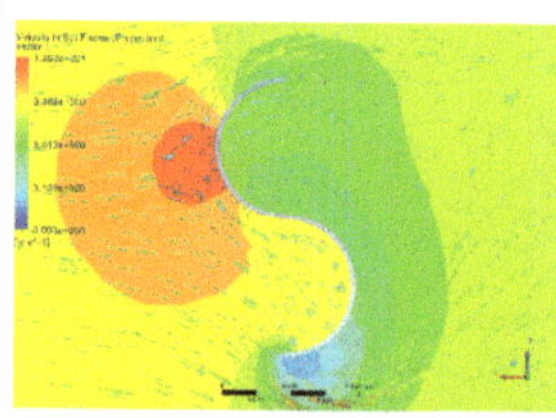	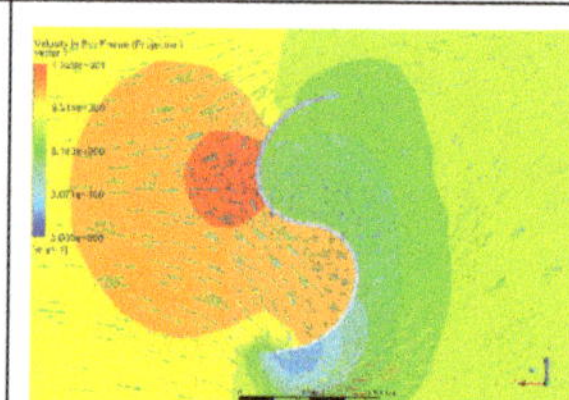
20°	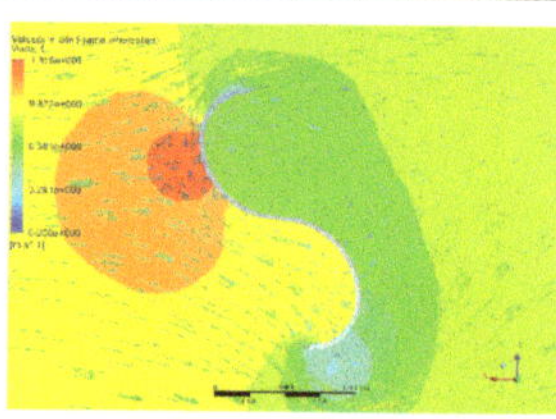	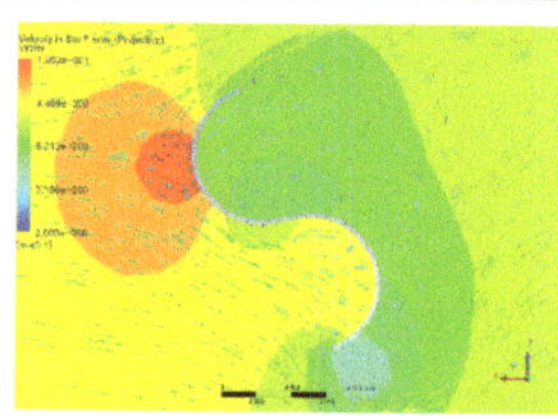	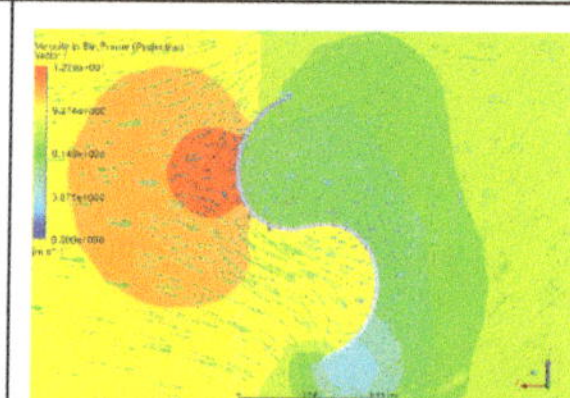
40°	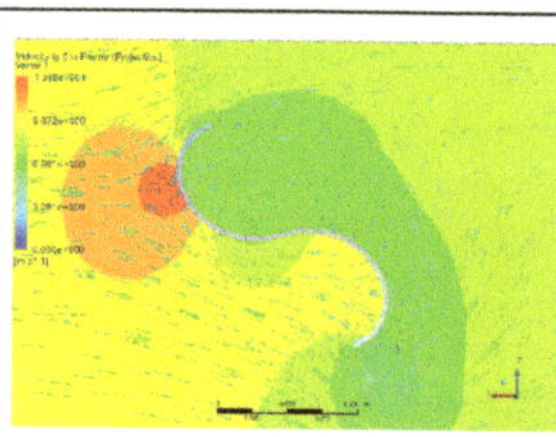	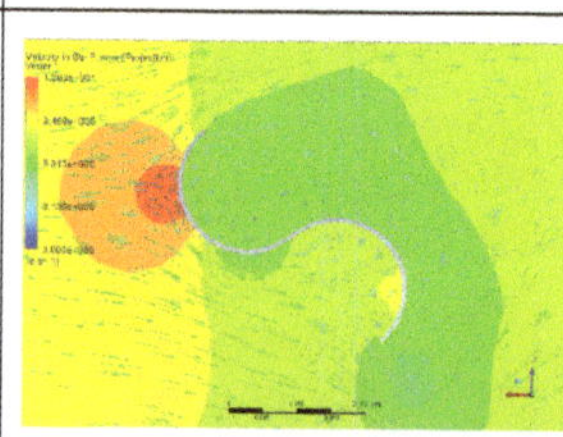	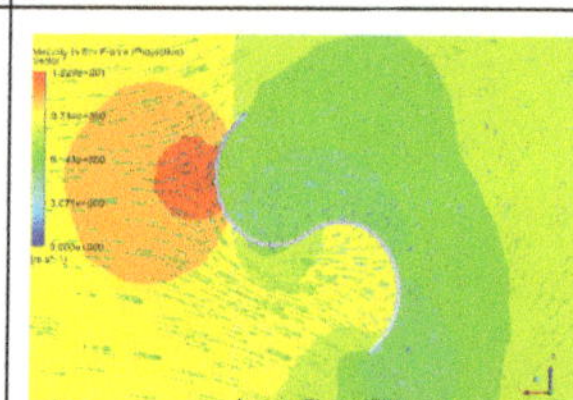

Figure 10. *Cont.*

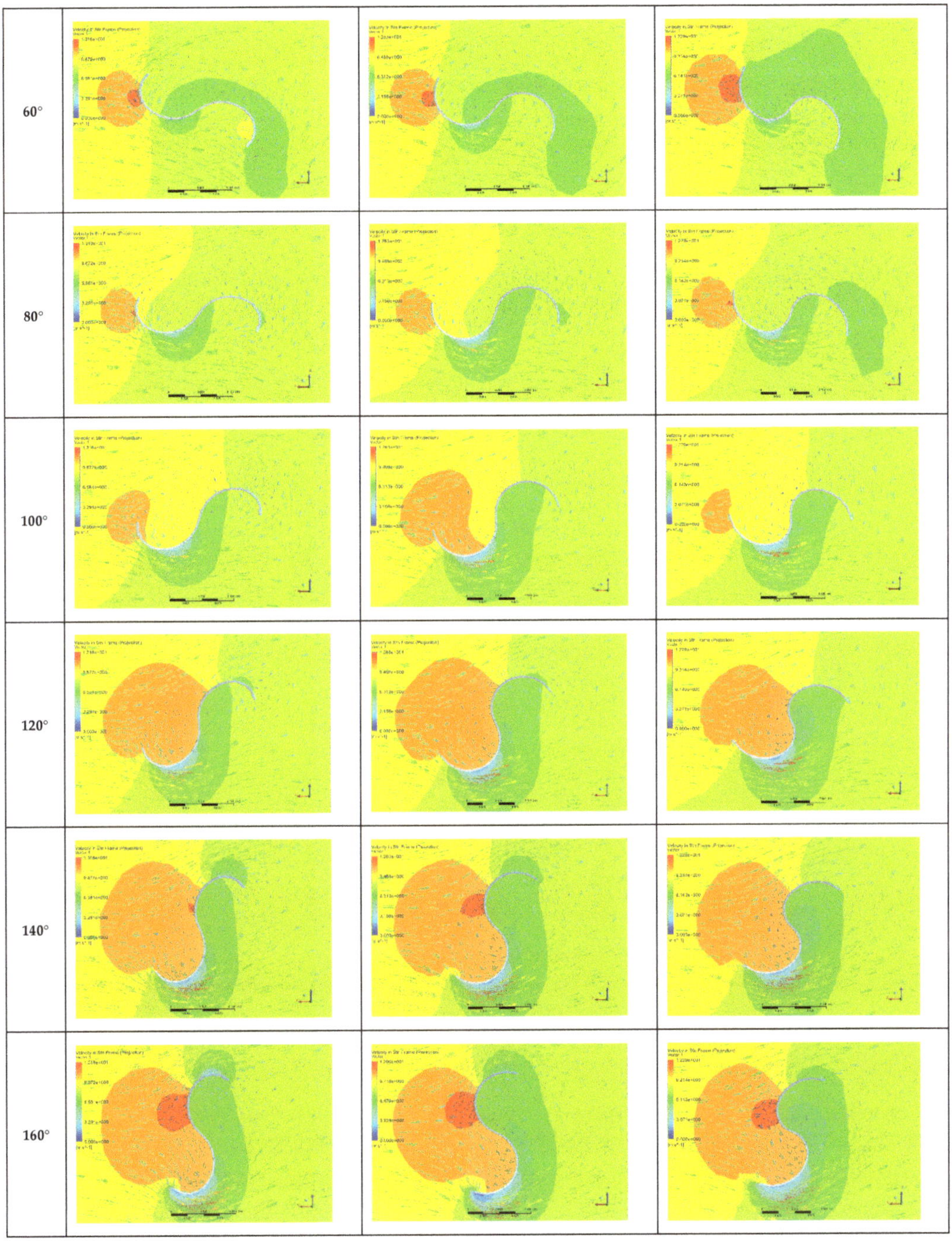

Figure 10. *Cont.*

180°

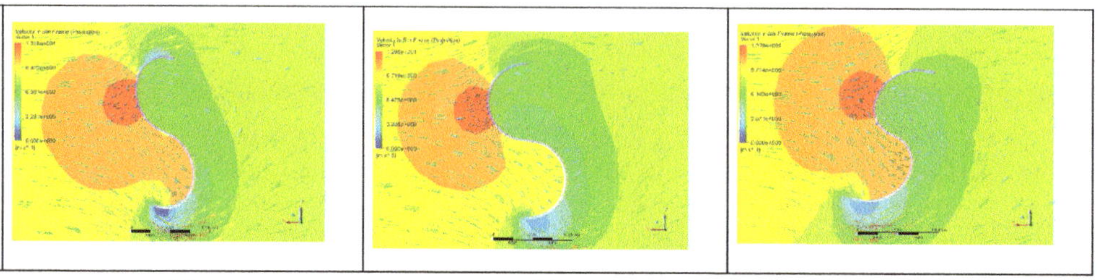

(**a**) (**b**) (**c**)

Figure 10. Velocity and pressure field for (**a**) Fibonacci I at TSR = 0.514 (ω = 37 rad/s); (**b**) Savonius at TSR = 0.507 (ω = 35 rad/s); (**c**) Fibonacci II at TSR = 0.506 (ω = 37 rad/s).

4. Conclusions

This paper proposes an improvement of the Savonius turbine which consists of implementing an innovative rotor geometry based on the Fibonacci spiral. These geometries were analyzed both experimentally and numerically. Two Fibonacci geometries were compared with the Savonius turbine, and it was found that the most efficient geometry is the one in which the lower radius arc is placed at the blade tip, Fibonacci I. This configuration provided higher power coefficients than the Savonius one. On the other hand, the Fibonacci II configuration lead to low power coefficients. This could be contradictory since, despite the fact that the Fibonacci II configuration is the one present in nature, Fibonacci I provided better results. This conclusion highlights the necessity of future works related to improving the performance of geometries based on the Fibonacci's spiral applied, for instance, to centrifugal pumps or fans. Some parameters that need to be analyzed are the separation gap (SG), overlap (OL), combination of SG and OL, and so on. It is worth mentioning that this is a very small model with an even smaller wind speed.

Author Contributions: Conceptualization, J.B. and J.d.D.R.; methodology, J.B. and J.d.D.R.; software, J.B. and M.I.L.; validation, J.B. and J.d.D.R.; formal analysis, J.B. and A.C.; investigation, J.B., J.d.D.R., A.C. and M.I.L.; resources, J.B., J.d.D.R., A.C. and M.I.L.; writing—original draft preparation, J.B., J.d.D.R. and M.I.L.; writing—review and editing, J.B., J.d.D.R., A.C. and M.I.L.; visualization, J.B., J.d.D.R., A.C. and M.I.L.; supervision, J.B., J.d.D.R., A.C. and M.I.L. All authors have read and agreed to the published version of the manuscript.

Funding: This research received no external funding.

Institutional Review Board Statement: Not applicable.

Informed Consent Statement: Not applicable.

Acknowledgments: The authors would like to express their gratitude to the Nautical Sciences and Marine Engineering department of the University of Coruña.

Conflicts of Interest: The authors declare no conflict of interest.

References

1. Hegler, S.; Plettemeier, D. Simulative investigation of the radar cross section of wind turbines. *Appl. Sci.* **2019**, *9*, 4024. [CrossRef]
2. Ding, L.; Guo, T. Numerical study on the power efficiency and flow characteristics of a new type of wind energy collection device. *Appl. Sci.* **2020**, *10*, 7438. [CrossRef]
3. Ancuti, M.C.; Musuroi, S.; Sorandaru, C.; Dordescu, M.; Erdodi, G.M. Wind turbines optimal operation at time variable wind speeds. *Appl. Sci.* **2020**, *10*, 4232. [CrossRef]
4. Dolinski, L.; Krawczuk, M. Analysis of modal parameters using a statistical approach for condition monitoring of the wind turbine blade. *Appl. Sci.* **2020**, *10*, 5878. [CrossRef]
5. Jessen, K.; Laugesen, K.; Mortensen, S.M.; Jensen, J.K.; Soltani, M.N. Experimental validation of aero-hydro-servo-elastic models of a scaled floating offshore wind turbine. *Appl. Sci.* **2019**, *9*, 1244. [CrossRef]
6. Liu, Z.; Tu, Y.; Wang, W.; Qian, G. Numerical analysis of a catenary mooring system attached by clump masses for improving the wave-resistance ability of a spar buoy-type floating offshore wind turbine. *Appl. Sci.* **2019**, *9*, 1075. [CrossRef]

7. Qi, L.; Zheng, L.; Bai, X.; Chen, Q.; Chen, J.; Chen, Y. Chen Nonlinear maximum power point tracking control method for wind turbines considering dynamics. *Appl. Sci.* **2020**, *10*, 811. [CrossRef]
8. Astolfi, D.; Castellani, F.; Berno, F.; Terzi, L. Numerical and experimental methods for the assessment of wind turbine control upgrades. *Appl. Sci.* **2018**, *8*, 2639. [CrossRef]
9. Han, X.; Liu, D.; Xu, C.; Shen, W.; Li, L.; Xue, F. Monin–Obukhov similarity theory for modeling of wind turbine wakes under atmospheric stable conditions: Breakdown and modifications. *Appl. Sci.* **2019**, *9*, 4256. [CrossRef]
10. Damota, J.; Lamas, I.; Couce, A.; Rodríguez, J. Vertical axis wind turbines: Current technologies and future trends. In Proceedings of the International Conference on Renewable Energies and Power Quality (ICREPQ'15), La Coruña, Spain, 25–27 March 2015.
11. Wijayanto, R.P.; Kono, T.; Kiwata, T. Performance characteristics of an orthopter-type vertical axis wind turbine in shear flows. *Appl. Sci.* **2020**, *10*, 1778. [CrossRef]
12. Pagnini, L.C.; Burlando, M.; Repetto, M.P. Experimental power curve of small-size wind turbines in turbulent urban environment. *Appl. Energy* **2015**, *154*, 112–121. [CrossRef]
13. Vita, G.; Šarkić-Glumac, A.; Hemida, H.; Salvadori, S.; Baniotopoulos, C. On the Wind Energy Resource above High-Rise Buildings. *Energies* **2020**, *13*, 3641. [CrossRef]
14. Li, J.; Yu, X.B. Analyses of the extensible blade in improving wind energy production at sites with low-class wind resource. *Energies* **2017**, *10*, 1295. [CrossRef]
15. Pujol, T.; Massaguer, A.; Massaguer, E.; Montoro, L.; Comamala, M. Net power coefficient of vertical and horizontal wind turbines with crossflow runners. *Energies* **2018**, *11*, 110. [CrossRef]
16. Tian, W.; Song, B.; VanZwieten, J.H.; Pyakurel, P. Computational fluid dynamics prediction of a modified Savonius wind turbine with novel blade shapes. *Energies* **2015**, *8*, 7915–7929. [CrossRef]
17. Savonius, S.J. The S-rotor and its applications. *Mech. Eng.* **1931**, *53*, 333–338.
18. Zhang, H.; Li, Z.; Xin, D.; Zhan, J. Improvement of aerodynamic performance of Savonius wind rotor using straight-arc curtain. *Appl. Sci.* **2020**, *10*, 7216. [CrossRef]
19. Vipin, G.K.; Ram, R.; Kumar, N. Numerical Analysis of different blade profile of wind turbine. *Int. J. Appl. Eng. Res.* **2018**, *13*, 375–385.
20. Alom, N.; Saha, U.K. Influence of blade profiles on Savonius rotor performance: Numerical simulation and experimental validation. *Energy Convers. Manag.* **2019**, *186*, 267–277. [CrossRef]
21. Kacprzak, K.; Liskiewicz, G.; Sobczak, K. Numerical investigation of conventional and modified Savonius wind turbines. *Renew. Energy* **2013**, *60*, 578–585. [CrossRef]
22. Benesh, A.H. Wind Turbine System Using a Savonius-Type Rotor. U.S. Patent 4,784,568, 29 September 1987.
23. Mohamed, M.H.; Janiga, G.; Pap, E.; Thevenin, D. Optimal blade shape of a modified Savonius turbine using an obstacle shielding the returning blade. *Energy Convers. Manag.* **2011**, *52*, 236–242. [CrossRef]
24. Chan, C.M.; Bai, H.L.; He, D.Q. Blade shape optimization of the Savonius wind turbine using a genetic algorithm. *Appl. Energy* **2018**, *213*, 148–157. [CrossRef]
25. Lamas, M.I.; Rodriguez, J.D.; Rodriguez, C.G. CFD analysis of biologically-inspired marine propulsors. *Brodogradnja* **2012**, *63*, 125–133.
26. Lamas Galdo, M.I.; Rodriguez Vidal, C.G.; Rodriguez García, J.D. Optimization of the efficiency of a biomimetic marine propulsor using CFD. *Ing. Investig.* **2014**, *34*, 17–21.
27. Lamas, M.I.; Rodríguez, J.D.; Rodríguez, C.G.; González, P.B. Three-dimensional CFD analysis to study the thrust and efficiency of a biologically inspired marine propulsor. *Pol. Marit. Res.* **2011**, *18*, 10–16. [CrossRef]
28. Eltayesh, A.; Hanna, M.B.; Castellani, F.; Huzayyin, A.S.; El-Batsh, H.M.; Burlando, M.; Becchetti, M. Effect of wind tunnel blockage on the performance of a horizontal axis wind turbine with different blade number. *Energies* **2019**, *12*, 1988. [CrossRef]
29. Versteeg, H.K.; Malalasekera, W. *An Introduction to Computational Fluid Dynamics: The Finite Volume Method*; Pearson Education Limited: London, UK, 2007; ISBN 978-0-13-127498-3.

applied sciences

MDPI

Article

Proposed Models to Improve Predicting the Operating Temperature of Different Photovoltaic Module Technologies under Various Climatic Conditions

Dang Phuc Nguyen Nguyen, Kristiaan Neyts and Johan Lauwaert *

Department of Electronics and Information Systems, Ghent University, Technologiepark Zwijnaarde 126, 9052 Ghent, Belgium; dangphucnguyen.nguyen@ugent.be (D.P.N.N.); kristiaan.neyts@ugent.be (K.N.)
* Correspondence: johan.lauwaert@ugent.be; Tel.: +32-9-264-6662

Abstract: The operating temperature is an essential parameter determining the performance of a photovoltaic (PV) module. Moreover, the estimation of the temperature in the absence of measurements is very complex, especially for outdoor conditions. Fortunately, several models with and without wind speed have been proposed to predict the outdoor operating temperature of a PV module. However, a problem for these models is that their accuracy decreases when the sampling interval is smaller due to the thermal inertia of the PV modules. In this paper, two models, one with wind speed and the other without wind speed, are proposed to improve the precision of estimating the operating temperature of outdoor PV modules. The innovative aspect of this study is two novel thermal models that consider the variation of solar irradiation over time and the thermal inertia of the PV module. The calculation is applied to different types of PV modules, including crystalline silicon, thin film as well as tandem technology at different locations. The models are compared to models that are described in the literature. The results obtained in different time steps show that our proposed models achieve better performance and can be applied to different PV technologies.

Keywords: photovoltaic; module temperature; PV operating temperature; module temperature models

Citation: Nguyen, D.P.N.; Neyts, K.; Lauwaert, J. Proposed Models to Improve Predicting the Operating Temperature of Different Photovoltaic Module Technologies under Various Climatic Conditions. *Appl. Sci.* **2021**, *11*, 7064. https://doi.org/10.3390/app11157064

Academic Editor: María Isabel Lamas Galdo

Received: 15 July 2021
Accepted: 27 July 2021
Published: 30 July 2021

Publisher's Note: MDPI stays neutral with regard to jurisdictional claims in published maps and institutional affiliations.

Copyright: © 2021 by the authors. Licensee MDPI, Basel, Switzerland. This article is an open access article distributed under the terms and conditions of the Creative Commons Attribution (CC BY) license (https://creativecommons.org/licenses/by/4.0/).

1. Introduction

Renewable energy, especially PV energy, has experienced strong growth in recent years. According to the International Energy Agency, solar PV generation increased by 131 TWh in 2019 and represented the second-largest generation growth of all renewable technologies, slightly behind wind energy [1]. With this increase, power generation from PV is approximately 720 TWh and shares almost 3% in global electricity generation. To optimize PV energy yield, besides inventing new technologies and reducing cost, predicting the performance of a PV module under on-site conditions plays an important role in selecting a PV module and installation method.

Various studies showed that the performance of a PV module depends not only on its own properties such as material, glazing-cover transmittance, and plate absorptance but also on the actual weather conditions such as ambient temperature, wind speed, and the spectral distribution of incident irradiance [2–5]. Even operating under conditions similar to Standard Test Conditions (STC) solar irradiance, the outdoor operating efficiency of multi-crystalline PV was found to be on average 18.1% lower than the given STC efficiency [6]. Typically, every 1 °C of increasing temperature results in a 0.5% reduction in efficiency of crystalline PV module, while this figure of amorphous silicon PV module is 0.27% on average [7].

One of the most important loss factors, which have a negative impact on the final energy produced by a PV system, is the operating temperature of the modules. This loss is represented by the temperature loss coefficient, which depends on the technology of the module as well as on the materials. Notwithstanding that those coefficient values are supplied by manufacturers, the cell operating temperature must be known as a prerequisite for estimating the total thermal loss. For this reason, accurately predicting the outdoor operating temperature plays an important role in modeling the energy yield of a PV system. As a result, various studies have been conducted to find the most suitable methodology for modeling the outdoor operating temperate of a PV module. According to Skoplaki and Palyvos, these models can be distributed in two main categories: implicit and explicit models [8].

The concept of implicit models is based on the knowledge of the thermal properties of the PV module and their heat transfer mechanisms, which is the so-called steady-state energy balance. These models have proven that they are able to determine the operating temperature of a PV module under outdoor conditions. However, this type of model seems complicated to implement in practice as they require the PV module to be in steady-state, which rarely happens under real operating conditions [9]. Moreover, since these models are composed of many factors that greatly depend on module materials and local meteorology, various parameters need to be provided at high precision to obtain the expected performance. This inconvenience makes it difficult to transfer the implicit models to other PV technologies since they have mainly been applied to crystalline silicon solar cells. One of the most well-known models is based on a simple energy balance proposed by Mattei et al. [10] and then developed by Akhsassi et al. [11].

The explicit models, on the other hand, emphasize the link between cell temperature and ambient temperature as well as the incident solar radiation flux. Some of them consider the wind speed, such as King et al. [3], Faiman et al. [12], and Koel et al. [13], while the others do not [11,14,15] but, in both cases, the solar irradiance is the main factor for increasing the PV module temperature. Moreover, the temperature of the module is strongly influenced by the thermal insulation of the module backside resulting from the roof-mounting or building integration [3,16]. Out of all models, the one put forward by King et al. [3] uses an exponential equation to describe the rising temperature caused by incident irradiance and the decrease in temperature caused by the on-site wind factor.

Other researchers refer to using nominal operating cell temperature (NOCT), which overlaps between implicit and explicit methods. The advantage of using NOCT is that this parameter is usually supplied by module manufacturers, and the implementation is simple. The disadvantage of this method is that the NOCT temperature is defined under specific meteorological conditions, which are difficult to meet under real conditions. Moreover, some studies showed that NOCT is not constant and varies by month, season, and location [9,17].

All those models listed above are able to predict the temperature values of the module back surface of the PV cells for both instant values and hourly time steps. However, Segado et al. [9] found that using hourly input parameters gained a higher accuracy of the prediction compared to using instant values of those parameters. Moreover, previous studies indicated that it is difficult to find a model, which can satisfy all PV module technologies [9,18,19], while Koehl et al. [13] showed that the accuracy of a thermal model is also affected by the on-site conditions.

Thus, one of the objects of this study is to improve the possibility of calculating PV module temperature by taking into account the effect of its own thermal inertia, which strongly impacts the changing of the module temperature, especially for locations where the solar irradiance and wind speed fluctuate strongly.

In this paper, two new module temperature models were proposed to predict the back-surface temperature of a PV module under outdoor operating conditions. The assessment considers parameters associated with the installation site, such as solar irradiance, ambient temperature, wind speed, mounting configuration, and interval recording, along with the PV cell material. The first model was achieved by modifying the method described by King, while the second model was based on the idea of the relationship between module temperature and solar irradiation intensity on the PV surface. A comparison between the proposed models and previous models existing in the literature was implemented to determine their reliability.

2. Materials and Methods

2.1. Data

Two databases were used for training and validating the proposed models. The first database was taken from ESMAP [20], which provide the measurements from different automated solar stations in Vietnam. Data for each site included one-minute averages of solar irradiance, PV module temperature, ambient temperature, and wind speed. Three data sets were selected from this database corresponding to three climatic zones in Vietnam: Hanoi in the north with subtropical with dry winter climate (Cwa), Danang in the center with equatorial monsoon climate (Am), and Tri An in the south with equatorial savannah with dry winter climate (Aw) [21]. The second database provides measured data for different PV modules in the USA, including crystalline silicon and thin film technologies, and is supplied by Marion [22]. Those modules were deployed in different locations: Cocoa, Florida, with subtropical climate; Eugene, Oregon, with marine west coast climate; and Golden, Colorado, with a semi-arid climate. Different from the former datasets, the data for these sites were measured immediately at a certain time of recording. The general characteristics of these databases are presented in Table 1.

Table 1. Selected databases for training and validating models.

Group	Location	Latitude (°N)	Interval Sampling (Minute)	Observed Period
1	Tri An [1]	11.01	1	01/2018–11/2019
	Da Nang [1]	16.01	1	01/2018–11/2019
	Ha Noi [1]	21.20	1	01/2018–09/2019
2	Cocoa [2]	28.39	5	01/2011–03/2012
	Golden [2]	39.74	15	08/2012–09/2013
	Eugene [2]	44.05	5	10/2012–01/2014

[1] Selected data from [20]. [2] Selected data from [22–24].

The data in Group 1 were used for both training and validating the proposed models, while the data in Group 2 were used for evaluating the accurate prediction of those models on different PV module technologies compared to other literature models. The PV module technologies and their characteristics used in this study are presented in Table 2. Due to the lack of wind measurements, which needs to be measured on-site at the same conditions as other parameters, measurements of the wind speed at the 10-m height [24] and the nearest measured station [23] were applied.

Table 2. Technologies and main characteristics of the different modules.

Module Technology	Characteristics		Sampling Points		
	η_{STC} (%)	$\beta_{P_{mp}}\left(\frac{\%}{^{\circ}C}\right)$	Cocoa	Golden	Eugene
Single-crystalline silicon (xSi) [1]	13.60	−0.42	29,632	10,266	37,820
Multi-crystalline silicon (mSi) [2]	14.00	−0.41	28,248	10,268	38,016
Cadmium telluride (CdTe) [1]	8.85	−0.21	29,730	10,362	37,062
Copper indium gallium selenide (CIGS) [1]	11.2	−0.39	29,714	10,405	37,815
Amorphous silicon/crystalline silicon (HIT) [1]	17.6	−0.35	29,119	10,292	37,985
Amorphous silicon/microcrystalline silicon (aSiMicro) [1]	7.5	−0.36	29,701	10,578	38,022
Amorphous silicon tandem junction (aSiTandem) [1]	4.47	−0.25	29,853	10,413	37,970

[1] Deployed in Cocoa, Golden and Eugene. [2] Deployed in Cocoa, Golden, Eugene, Tri An, Da Nang and Ha Noi.

2.2. Previous Models

There are many temperature models in the literature, so we have implemented and compared seven existing models presented in Table 3 with our proposed models in this study. In these selected models, some of them are able to predict the module temperature while others calculate the cell temperature directly. For the models predicting cell temperature, which becomes significantly different from the back-surface module temperature in high solar radiation intensities [3], the following relationship was applied:

$$T_c = T_m + \frac{G_g}{G_0} \cdot \Delta T_m, \quad (1)$$

Table 3. Literature models for validating proposed models.

Groups	Correlations	Comments	Ref.	Equation
With wind	Sandia: $T_m = T_a + G_g \exp(a + bW_s)$	$a = -3.56$ $b = -0.075$ s/m	[3]	(2)
	Faiman: $T_m = T_a + \frac{G_g}{U_0 + U_1 W_s}$	$U_0 = 30.2$ W/m²/°C $U_1 = 6.28$ Ws/m³/°C	[12,13]	(3)
	PVSyst1: $T_c = T_a + G_g \frac{\alpha(1-\eta_m)}{U_0 + U_1 W_s}$	$U_0 = 25$ W/m²/°C $U_1 = 1.2$ Ws/m³/°C $\alpha = 0.9$, $\eta_m = 0.1$	[15]	(4)
	Akhsassi1: $T_m = \frac{U_L T_a + \left[(\tau\alpha) - \eta_{STC}\left(1 - \beta_{P_{mp}} T_0\right)\left(1 + \gamma_{P_{mp}} Ln\left(\frac{G_g}{G_0}\right)\right)\right] G_g}{U_L + \eta_{STC} \beta_{P_{mp}} \left(1 + \gamma_{P_{mp}} Ln\left(\frac{G_g}{G_0}\right)\right) G_g}$	$U_L = U_0 + U_1 W_s$ $U_0 = 24.68$ W/m²/°C $U_1 = 6.13$ Ws/m³/°C	[11]	(5)
Without wind	Lasnier: $T_m = 30\ °C + 0.0175\ °C \frac{m^2}{W}\left(G_g - 300 \frac{W}{m^2}\right) + 1.14(T_a - 25\ °C)$		[14]	(6)
	PVSyst2: $T_c = T_a + G_g \frac{\alpha(1-\eta_m)}{U_0 + U_1 W_s}$	$U_0 = 29$ W/m²/°C $U_1 = 0$ Ws/m³/°C $\alpha = 0.9$, $\eta_m = 0.1$	[15]	(7)
	Akhsassi2: $T_m = 25\ °C + 0.123\ °C \frac{m^2}{W}\left(G_g - 200\ W/m^2\right) + 1.04(T_a - 20\ °C)$		[11]	(8)

The temperature difference between the PV cell and module back surface was evaluated by King et al. [3] to be about ΔT_m = 3 °C at an irradiance level of $G_g = G_0$ for an open-rack installation PV system.

For the first Akhsassi model, the module efficiency and temperature coefficient of maximum power at STC were taken directly from the PVs' characteristics. While the solar irradiance coefficient $\gamma_{P_{mp}}$ is between 0.03 and 1.12 for silicon and most often simplified to $\gamma_{P_{mp}} = 0$ [10,11,25], it was chosen to be 0.04 in this study.

2.3. Selected Metrics for Evaluating the Methods

In order to evaluate the differences between the measured ($T_{measured}$) and predicted ($T_{predicted}$) operating temperature of the PV modules, three statistical metrics have been applied:

- The correlation coefficient R^2 defines the relationship between the estimated and measured data as the following expression:

$$R^2 = \frac{\left(\sum_i \left[\left(T_{predicted} - \overline{T}_{predicted}\right)\left(T_{measured} - \overline{T}_{measured}\right)\right]\right)^2}{\sum_i \left(T_{predicted} - \overline{T}_{predicted}\right)^2 \left(T_{measured} - \overline{T}_{measured}\right)^2}, \quad (9)$$

- The root mean square error, used to evaluate the fluctuations around the model and defined by the expression:

$$RMSE = \sqrt{\frac{\sum_i^n \left(T_{measured} - T_{predicted}\right)^2}{n}}, \quad (10)$$

2.4. Proposed Models

2.4.1. Model with Wind

As mentioned above, several studies evidently showed that there is no best option, as we could not apply a unique temperature model for all PV module technologies. However, it was clear that the Sandia (King) model is frequently on the top list of matching temperature models. Besides, the Sandia model, unlike other temperature models, only takes on-site parameters such as plane-of-array (POA) irradiance, wind speed, and installation method into consideration. This reduction of input parameters leads to a simple implementation since it does not require many input parameters and a deep understanding of module technologies and materials.

As PV thermal models usually describe the module temperature based on the assumption of thermal equilibrium, the thermal mass of the module has a high impact on the responding time of the changes in the temperature [26]. Moreover, Segado et al. proved that the temperature prediction of a PV module is strongly affected by its own thermal inertia [9]. However, literature studies showed that Sandia is one of the models that is least affected by meteorological conditions. In other words, the changes in its empirical efficiencies are insignificant under different weather conditions and PV technologies [9,18,26]. Nevertheless, it was recognized that this model usually performed well under the low intensity of POA irradiance and less fluctuation of weather conditions; otherwise, it could suffer performance loss in practical implementation.

In this research, therefore, in order to improve the calculation of the module back-surface temperature, a modification of the Sandia model (Equation (2)) by adding a correction term was put forward, based on the idea of the relevance between module thermal inertia and interval sampling, presented by the following Equation:

$$T_m = T_a + \Delta T_{ref}^{*} \frac{G_g}{G_0} Exp\left(-\frac{W_s}{W_{ref}}\right) - \Delta T_{ref} \frac{\Delta G_g}{G_0} f(\Delta t), \quad (11)$$

where $\Delta G_g(t) = \left[G_g(t) - G_g(t-1)\right]$ is a dimensional constant describing the difference of POA irradiance between two consecutive sampling points ($\Delta G_g = 0\ \mathrm{W/m^2}$ at $t = 0$); Δt is the time recording interval in minutes; and ΔT_{ref}^{*} and W_{ref} are empirical coefficients depending on module installation configuration and are equivalent to the coefficients a an b in the Sandia model from Ref. [3] and Equation (2). With $\Delta T_{ref}^{*} = G_0 e^a$ and $W_{ref} = -\frac{1}{b}$, this correction term becomes exactly the wind correction of Equation (2).

2.4.2. Model without Wind

As wind data are not always available for every PV system, there were many models without wind speed were presented to estimate the operating temperature of a PV cell/module. Some of them took into account the manufacturing and installing characteristics [15,27], while others neglected those parameters [11,14]. In this paper, we proposed a model based on the proportion of the POA irradiance to the solar irradiance at STC to calculate the operating temperature of a PV module without using wind velocity. At this suggestion, we assumed that the changing temperature of a PV module is proportional to the varying POA irradiance at operating conditions compared to the laboratory conditions.

To improve the accuracy, the thermal mass of the PV module is once again taken into account, and our proposed model is expressed as the following Equation:

$$T_m = T_a + \Delta T^*_{ref}\frac{G_g}{G_0} - \Delta T_{ref}\frac{\Delta G_g}{G_0} f(\Delta t), \tag{12}$$

3. Results and Discussion

3.1. Parametric Identification

The measurements in 2018 from Group 1 in Table 1, which include POA irradiance, air temperature, wind speed, and module temperature, were selected to determine the coefficients of the suggested models. In these locations, the multi-crystalline silicon modules were installed with fixed open cracks; therefore, the coefficients ΔT^*_{ref} and W_{ref} in Equation (11) were taken as 28.4 °C and 13.3 m/s, respectively. These values correspond with $a = -3.56$ and $b = -0.075$ s/m in the original Sandia model [3].

In the next stage, the best fit for the values $y = \frac{\Delta T_{ref}}{G_0} f(\Delta t)$, describing the relevance of the thermal inertia of the module and the interval sampling, were obtained as a function of the time step Δt for both models, based on measured data in three cities in 2018. The measurement data are available for every minute, and to determine y based on the time step Δt, the measurement data of T_m and ΔG_g are averaged over the duration of the time step Δt. The time step is limited to 10 min because otherwise, the time dependency is not described in sufficient detail. The resulting values of y and the corresponding correlation (R^2) and error ($RMSE$) between the simulated values of the temperature T_m and the measurement data for the proposed model with wind speed are presented in Table 4. The mean $\bar{y}$, averaged over the three locations, and the related standard deviation (SD) were also calculated. The obtained values of y as a function of Δt for both proposed models are illustrated in Figure 1. The relation between $\bar{y}$ and Δt is approximately exponential and a least-squares fit yields R^2 equal to 0.97 and 0.99 for the model with wind speed and the model without wind speed, respectively: $\bar{y} = 0.0175\exp(-0.061\Delta t)$ and $\bar{y} = 0.0162\exp(-0.056\Delta t)$. These relations can be used to eliminate the function $f(\Delta t)$ that we have introduced previously.

Table 4. Calculated values of $y = \frac{\Delta T_{ref}}{G_0} f(\Delta t)$, correlation ($R^2$), and error ($RMSE$) between measured and simulated temperature T_m for three cities and for different values of the time step Δt.

Time Step (Min.)	Location									Average	
	Tri An			Da Nang			Ha Noi				
	$y\,(\frac{°C\,m^2}{W})$	R^2 (%)	$RMSE$ (°C)	$y\,(\frac{°C\,m^2}{W})$	R^2 (%)	$RMSE$ (°C)	$y\,(\frac{°C\,m^2}{W})$	R^2 (%)	$RMSE$ (°C)	$\bar{y}\,(\frac{°C\,m^2}{W})$	SD
1	0.0153	91.4	2.81	0.0150	93.6	2.73	0.0166	93.7	3.83	0.0156	0.00085
2	0.0151	93.5	2.57	0.0148	95.2	2.51	0.0165	95.1	3.66	0.0155	0.00091
3	0.0144	94.6	2.43	0.0143	96.2	2.37	0.0161	95.9	3.56	0.0149	0.00101
4	0.0133	95.2	2.35	0.0135	96.7	2.29	0.0155	96.4	3.50	0.0141	0.00122
5	0.0123	95.5	2.32	0.0129	96.7	2.24	0.0149	96.7	3.46	0.0134	0.00136
6	0.0114	95.7	2.30	0.0121	97.2	2.21	0.0142	96.9	3.45	0.0126	0.00146
7	0.0101	95.7	2.30	0.0110	97.2	2.20	0.0128	96.9	3.45	0.0113	0.00137
8	0.0093	95.7	2.30	0.0103	97.3	2.19	0.0121	97.0	3.44	0.0106	0.00142
9	0.0083	95.7	2.31	0.0093	97.3	2.19	0.0113	97.1	3.45	0.0096	0.00153
10	0.0084	95.8	2.29	0.0095	97.4	2.17	0.0113	97.2	3.44	0.0097	0.00146

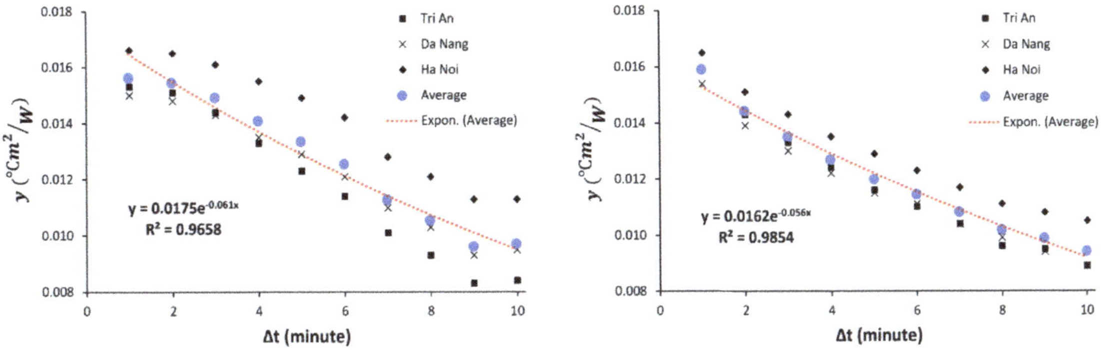

(a) (b)

Figure 1. Plot of the values of $y = \frac{\Delta T_{ref}}{G_0} f(\Delta t)$ in different time steps from Table 4 and exponential fitting curve for the mean value $\bar{y}$. (**a**) Model with wind speed. (**b**) Model without wind speed.

The Equation (11), therefore, becomes:

$$T_m = T_a + \Delta T^*_{ref} \frac{G_g}{G_0} \exp\left(-\frac{W_s}{W_{ref}}\right) - \Delta T_{ref} \frac{\Delta G_g}{G_0} \exp\left(-\frac{\Delta t}{\tau}\right), \quad (13)$$

where $\Delta T_{ref} = 18\ ^{\circ}\mathrm{C}$ and $\tau = 16.7$ min.

The same method was applied for the second model and then the Equation (12) becomes:

$$T_m = T_a + \Delta T^*_{ref} \frac{G_g}{G_0} - \Delta T_{ref} \frac{\Delta G_g}{G_0} \exp\left(-\frac{\Delta t}{\tau}\right), \quad (14)$$

with $\Delta T^*_{ref} = T_0 = 25\ ^{\circ}\mathrm{C}$, $\Delta T_{ref} = 16\ ^{\circ}\mathrm{C}$ and $\tau = 16.7$ min.

3.2. Models Comparison

At this stage, the recorded data in 2019 from three locations in Vietnam (Group 1 in Table 1) were taken out for evaluating the proposed models. After removing the samples which were recorded in the nighttime and all incorrect intervals, the total number of sampling points and corresponding time steps are presented in Table 5.

Table 5. The sample size of measurements in different time steps for three regions in Vietnam.

Time Step (Minute)	Sampling Point		
	Tri An	Da Nang	Ha Noi
1	246,982	247,703	198,476
5	49,401	49,542	39,703
10	24,395	24,467	19,597
15	16,136	16,187	12,975

Besides comparing experimental measurements and model predictions, a comparison between proposed and literature models was also implemented and presented in Table A1 (see Appendix A). As was to be expected, our proposed model with wind presented the best correlation between calculations and measurements with R^2 greater than 95% compared to other models, while the without-wind proposed model was the best in the group of models without wind speed. Moreover, the proposed model with wind illustrates that it has the most stable performance compared to all other models from the literature, as it is the only one that provided *RMSE* values less than 3 °C for all regions. In contrast, the *RMSE* values for other models changed significantly between the three locations. Moreover, this method also obtained better results for both R^2 and $RMSE$ coefficients at all time steps compared to the original Sandia model.

It was clear that the accuracy of all models was proportional to the interval time. As mentioned above, these models describe the temperature of PV modules under the assumption of their thermal equilibrium. Therefore, if this condition is not satisfied, then large differences can result. Moreover, it is known that the time response to changes in the temperature depends on the thermal mass of the module and the prevailing conditions [26], so that if the time resolution of the data is shorter than the typical thermal response times, the performance of the models will not be as good.

The difference between the predicted and measured values of the PV module back-surface temperature, ΔT, is illustrated in Figure 2. As we can see in the picture, the calculation for Sandia is proportional to the magnitude of fluctuating POA between two consecutive sampling points. The larger the variation, the greater the difference is. While the corresponding values for our proposed models tend to approach zero and are evenly distributed around zero across the range of ΔG_g.

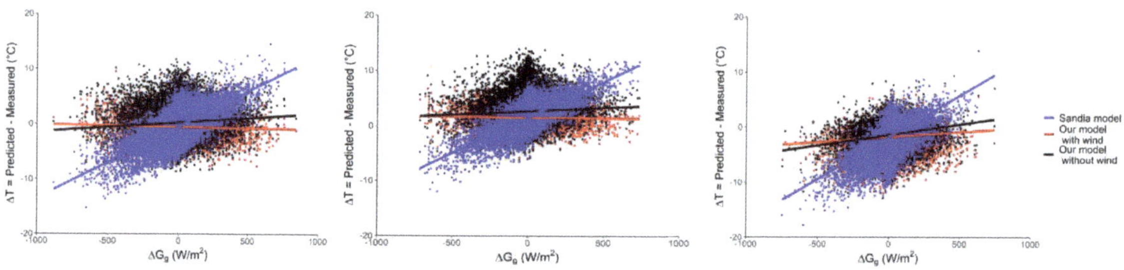

(a) (b) (c)

Figure 2. A comparison of the difference between predicted and measured temperature of PV module in three different locations, resulting in changing of incident POA with 5-min time steps: (**a**) Tri An; (**b**) Da Nang; and (**c**) Ha Noi.

We also took into account the implementation of temperature models at different POA levels, which strongly affected the accuracy of the predicting process. The results displayed in Figure 3 demonstrate that the accuracy of the temperature models will decrease when the radiation intensity reaches the module surface increase. This circumstance occurred because the fluctuations among high POA values happen more frequently than those of low POA levels. Moreover, the greater POA the module gets, the higher the temperature of the module will be. Subsequently, the thermal inertia of the PV modules will have a stronger effect when POA increases, which explains why our proposed models perform better than literature models at high POA levels but remain equivalent when POA levels are low.

Figure 4 showed that the on-site wind speed of all regions remained stable during the observed period, while the average POA irradiance occurred in different ways. The POA irradiance in Tri An is distributed across the whole range of values, while the values of that parameter in Ha Noi are mainly concentrated below 500 W/m^2. Moreover, Figure 4c illustrates that the distribution of the difference between two consecutive sampling points of wind is almost the same for all regions with variations between 0 and 3 m/s. While the largest fluctuation of POA irradiance occurred in Tri An, and the smallest fluctuation of that happened in Ha Noi. This means that the effect of thermal inertia on the thermal prediction in Tri An is greater than for the other regions due to its high intensity and frequent fluctuation of POA irradiance. Consequently, our proposed model with wind obtained the best result in Tri An where the on-site conditions fluctuated most significantly, while the proposed model without wind gained better performances in Tri An and Ha Noi, where the wind was less important than in Da Nang.

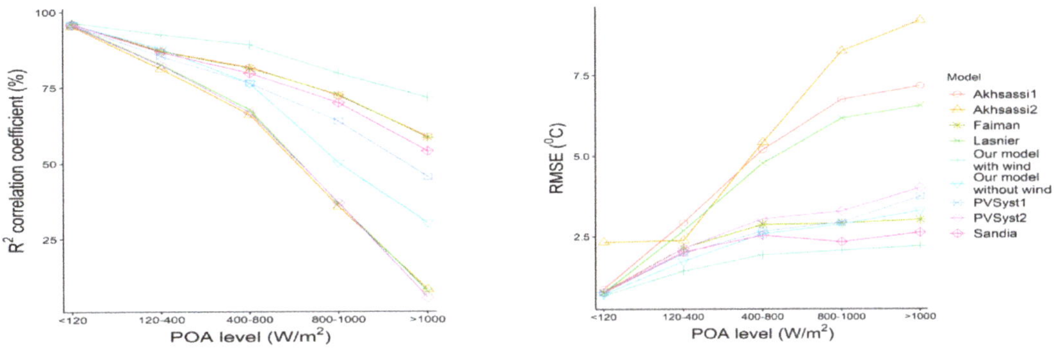

(a) (b)

Figure 3. Correlation between measured and calculated module temperature for different POA levels obtained with the studied models for Tri An, Vietnam (time recording = 5 min): (**a**) correlation factor R^2; (**b**) $RMSE$.

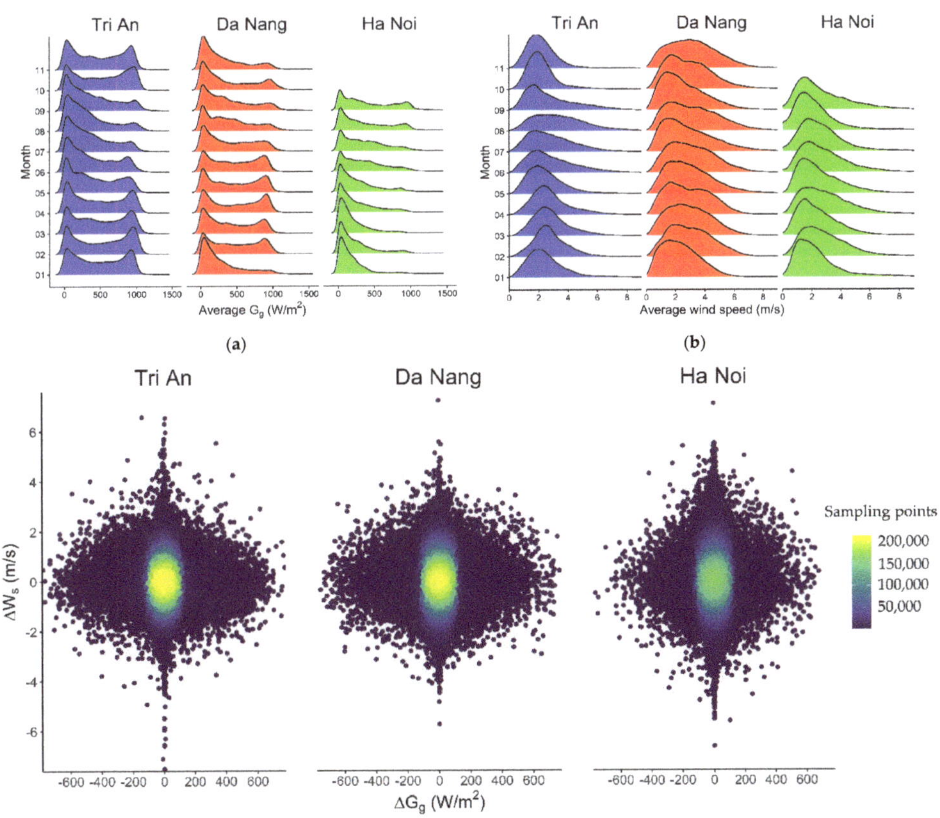

(c)

Figure 4. The distribution of POA irradiation and wind in three locations in Vietnam (1-min time step): (**a**) average monthly distribution of POA irradiation level; (**b**) average monthly distribution of wind speed; and (**c**) the variation of POA irradiance and wind speed between two consecutive sampling points.

The back-surface temperature of the PV module that is calculated for three regions by our proposed model with wind effect (Equation (13)) using optimized values of ΔT^*_{ref} and W_{ref} coefficients is presented in Table 6. By applying these best-fitted coefficients, the obtained $RMSE$ values were decreased significantly to below 2 °C for all regions.

Table 6. Correlation coefficients (R^2 and $RMSE$) for measured and calculated module temperatures for three regions in Vietnam using optimal coefficients (5-min time step).

Coefficients	Implementing Results		
	Tri An	Da Nang	Ha Noi
ΔT^*_{ref} (°C)	31.4	24.7	34.7
W_{ref} (m/s)	11.1	11.8	12.8
R^2 (%)	98.0	98.4	98.4
$RMSE$ (°C)	1.43	1.40	1.55

3.3. Comparison between Different PV Technologies

Our proposed models were examined afterwards with different module technologies, which were deployed in three locations in the USA. The obtained results for all regions are shown in Table A2 (see Appendix A), while a visual presentation for Cocoa is depicted in Figure 5. As we can see, our proposed model with wind obtained the best compatibility of predicting module temperature since it showed the highest correlation coefficient R^2 and the lowest $RMSE$, whilst the other proposed model has the best performance among the models that ignore the influence of the wind, with highest R^2 and lowest $RMSE$ values calculated for all PV technologies and locations.

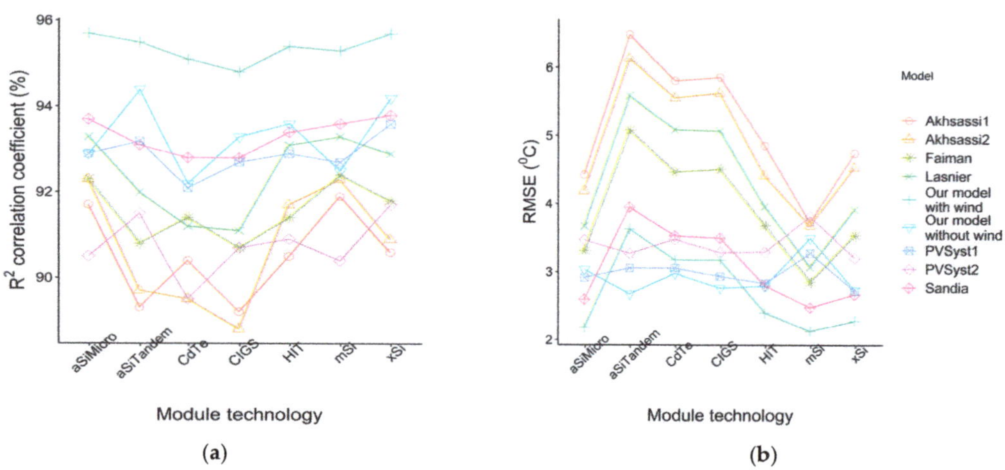

Figure 5. Correlation between measured and calculated module temperature for different module technologies obtained with the studied models for Cocoa, USA: (**a**) correlation factor R^2; (**b**) $RMSE$.

Previous studies have reported that it is difficult to apply a single model or a unique formula to precisely calculate the PV module/cell temperature [9,11,18,19]. Moreover, the thermal characteristics of PV modules are slightly different even if they are manufactured with the same technology and materials [12,13]. However, our proposed models have demonstrated that they can work well with different PV technologies and weather conditions without adjusting the empirical coefficients.

Table 7 shows the optimal coefficients, ΔT^*_{ref} and W_{ref} in Equation (13), and corresponding implementing results which were calculated for various PV technologies in three different locations. As we can see, the operating temperature of CdTe and CIGS modules were least affected by on-site POA, with ΔT^{*}_{ref} is 31.25 ± 0.45 °C and 30.65 ± 0.75 °C, respectively. Silicon PV module operating temperatures, on the other hand, were influenced significantly by local POA irradiance conditions, with ΔT^{*}_{ref} equal to 28.6 ± 1.3 °C and 27.65 ± 1.95 °C, respectively, for xSi and mSi modules.

Table 7. Correlation coefficient R^2 and $RMSE$ between measured and calculated module temperatures for different PV technologies using optimal coefficients.

PV Technology	Implementing Results											
	Cocoa				Golden				Eugenea			
	ΔT^{*}_{ref} (°C)	W_{ref} (m/s)	R_2 (%)	$RMSE$ (°C)	ΔT^{*}_{ref} (°C)	W_{ref} (m/s)	R_2 (%)	$RMSE$ (°C)	ΔT^{*}_{ref} (°C)	W_{ref} (m/s)	R_2 (%)	$RMSE$ (°C)
xSi	27.3	21.7	96.2	2.00	29.9	12.2	97.0	2.86	29.6	8.3	97.4	2.04
mSi	25.7	18.5	95.5	2.05	29.6	18.9	97.3	2.54	27.3	9.7	97.7	1.86
CdTe	30.8	17.5	94.9	2.47	31.7	15.2	97.0	2.54	30.8	8.5	96.9	2.37
CIGS	29.9	20.8	95.2	2.37	31.4	10.1	96.3	2.93	30.2	8.3	96.4	2.41
HIT	27.3	22.2	95.8	2.10	29.6	21.3	96.5	3.02	27.9	8.3	97.5	2.02
aSiMicro	27.1	19.6	95.8	2.06	29.6	18.9	96.6	2.90	29.0	8.5	97.1	2.19
aSiTandem	31.1	22.2	95.9	2.27	29.6	19.2	96.5	2.91	31.7	8.8	97.0	2.38

4. Conclusions

As the temperature has a significant effect on the electrical efficiency of solar cells, it is necessary to accurately estimate the photovoltaic module/cell temperature when predicting the energy yield. In this framework, we proposed two new models in order to calculate the PV module temperature. One model with the wind was developed based on the Sandia model by considering the effect of the thermal inertia of the module, whilst the other was determined without taking into account the wind speed. These proposed models have been compared to other models in the literature by using the statistical coefficients R^2 and $RMSE$. The results show that the inclusion of the thermal inertia of the PV module and the wind speed with the given models allow making a more accurate estimation of the module temperature, with R^2 correlation above 95% and the $RMSE$ below 3 °C for most PV technologies.

As they do not need many input parameters and are simple to implement, these proposed models have demonstrated that they also work well under different environmental and operating conditions and PV technologies. The proposed model with wind, Equation (13), can be used to calculate PV module temperature with high precision for any location where wind data are available; otherwise, Equation (14) can be applied. These thermal models are useful for evaluating the thermal performance of a PV module under different environmental and operating conditions since they have proven their reliability when being examined in different locations and climates.

When climatological information such as wind speed and solar illumination are available in a time series for a certain location, our models make it possible to estimate the module temperature for different types of solar cells. The temperature information is essential in estimating the PV module efficiency and in deciding which kind of PV panel will be most economical for the given location.

Author Contributions: Conceptualization, D.P.N.N. and J.L.; Data curation, D.P.N.N.; Formal analysis, D.P.N.N.; Funding acquisition, J.L.; Investigation, D.P.N.N.; Methodology, D.P.N.N., K.N. and J.L.; Project administration, J.L.; Resources, K.N. and J.L.; Software, D.P.N.N.; Supervision, K.N. and J.L.; Validation, K.N. and J.L.; Writing—original draft, D.P.N.N.; Writing—review and editing, K.N. and J.L. All authors have read and agreed to the published version of the manuscript.

Funding: This research received no external funding.

Institutional Review Board Statement: Not applicable.

Informed Consent Statement: Not applicable.

Data availability Statement: Data available on request.

Conflicts of Interest: The authors declare no conflict of interest.

Nomenclature

T_m	Module temperature (°C)
T_c	Cell temperature (°C)
T_a	Ambient temperature (°C)
T_0	Reference temperature of 25 °C
η_{STC}	Module efficiency coefficient (%)
$\beta_{P_{mp}}$	Module temperature coefficient (%/°C)
G_g	Plane Of Array irradiance (POA) (W/m^2)
G_0	Reference solar irradiance of 1000 W/m^2
W_s	Wind speed (m/s)
U_0	The constant heat transfer component (W/m^{2}K)
U_1	The convective heat transfer component (W/m^{3}sK)
$\gamma_{P_{mp}}$	Dimensionless coefficient between 0.03 and 0.12

Appendix A

Table A1. Results obtained for predicting module temperature with different time steps and models.

Location	Model	R^2 (%)				*RMSE* (°C)			
		Time Step (Minutes)							
		1	5	10	15	1	5	10	15
Tri An	Sandia *	94.3	96.1	97.0	97.5	2.45	2.06	1.84	1.72
	Faiman *	94.4	96.5	97.5	98.0	2.69	2.35	2.17	2.09
	PVSyst1 *	93.6	95.4	96.3	96.7	2.69	2.27	2.03	1.90
	Akhsassi1 *	94.5	96.5	97.4	97.8	4.55	4.46	4.42	4.41
	Our model with wind *	**95.6**	**97.8**	**98.1**	**98.1**	**2.18**	**1.63**	**1.56**	**1.55**
	Lasnier	92.5	93.8	94.4	94.7	4.19	4.08	4.03	4.01
	PVSyst2	92.0	93.7	94.6	95.0	2.88	2.53	2.33	2.22
	Akhsassi2	91.4	92.5	93.0	93.2	5.12	5.08	5.06	5.05
	Our model without wind	93.3	95.5	95.7	95.7	2.63	2.15	2.09	2.07
Da Nang	Sandia *	95.9	97.3	97.9	98.2	2.52	2.22	2.09	2.01
	Faiman *	95.2	96.9	97.6	98.0	**2.20**	**1.82**	**1.64**	**1.55**
	PVSyst1 *	95.4	96.7	97.3	97.6	3.63	3.41	3.31	3.26
	Akhsassi1 *	95.1	96.6	97.2	97.5	2.75	2.61	2.55	2.52
	Our model with wind *	**96.7**	**98.5**	**98.7**	**98.8**	2.35	1.95	1.88	1.87
	Lasnier	94.6	95.4	95.8	96.0	2.21	2.05	1.96	1.92
	PVSyst2	94.1	95.3	95.8	96.1	3.82	3.63	3.54	3.50
	Akhsassi2	93.6	94.2	94.5	94.6	3.21	3.14	3.10	3.07
	Our model without wind	95.0	96.4	96.5	96.5	3.72	3.49	3.46	3.46
Ha Noi	Sandia *	96.3	97.2	97.7	97.9	3.06	2.89	2.80	2.75
	Faiman *	95.8	96.9	97.5	97.8	3.39	3.24	3.16	3.13
	PVSyst1 *	96.2	97.1	97.7	98.0	**2.60**	**2.36**	**2.22**	**2.14**
	Akhsassi1 *	94.9	95.8	96.3	96.5	4.91	4.89	4.87	4.87
	Our model with wind *	**96.9**	**98.2**	**98.4**	**98.5**	2.94	2.66	2.63	2.63
	Lasnier	93.7	94.2	94.5	94.7	4.84	4.80	4.78	4.77
	PVSyst2	95.6	96.5	97.0	97.2	2.94	2.77	2.66	2.60
	Akhsassi2	91.4	91.7	91.9	92.1	5.14	5.12	5.11	5.11
	Our model without wind	96.3	97.5	97.7	97.8	2.79	2.49	2.45	2.45

* Models with wind speed.

Table A2. The statistical coefficients (R^2 and $RMSE$) for all models with and without wind speed vs. experimental measurement of module back-surface temperature calculated for different technologies.

Coefficient	Location	Module	Model								
			With Wind					Without Wind			
			Sandia	Faiman	PVSyst1	Akhsassi1	Our Model with Wind	Lasnier	PVSyst2	Akhsassi2	Our Model without Wind
R^2 (%)	Cocoa	xSi	93.8	91.8	93.6	90.6	**95.7**	92.9	91.7	90.9	94.2
		mSi	93.6	92.4	92.7	91.9	**95.3**	93.3	90.4	92.3	92.5
		CdTe	92.8	91.4	92.1	90.4	**95.1**	91.2	89.5	89.5	92.2
		CIGS	92.8	90.7	92.7	89.2	**94.8**	91.1	90.7	88.8	93.3
		HIT	93.4	91.4	92.9	90.5	**95.4**	93.1	90.9	91.7	93.6
		aSiMicro	93.7	92.3	92.9	91.7	**95.7**	93.3	90.5	92.3	92.9
		aSiTandem	93.1	90.8	93.2	89.3	**95.5**	92.0	91.5	89.7	94.4
	Golden	xSi	96.1	95.2	96.2	93.6	**96.8**	91.4	95.1	87.5	95.8
		mSi	95.9	94.5	96.3	92.8	**96.8**	91.7	95.6	87.7	96.5
		CdTe	95.4	94.5	95.6	92.8	**96.5**	90.9	94.7	87.1	95.7
		CIGS	95.0	94.5	94.8	93.1	**96.1**	90.3	93.4	86.8	94.4
		HIT	94.8	93.5	95.2	91.9	**96.0**	91.2	94.7	87.4	95.9
		aSiMicro	95.0	93.7	95.4	92.0	**96.1**	91.0	94.9	87.1	95.9
		aSiTandem	94.7	93.3	95.2	91.5	**95.9**	90.2	94.6	86.2	95.8
	Eugene	xSi	96.5	96.5	96.3	96.3	**97.3**	93.6	96.0	90.5	96.7
		mSi	96.8	96.6	96.5	96.8	**97.6**	95.5	96.5	93.0	97.3
		CdTe	95.9	95.8	95.6	95.8	**96.8**	93.6	95.4	90.8	96.3
		CIGS	95.5	95.5	95.2	95.4	**96.4**	92.9	95.0	90.1	95.8
		HIT	96.4	96.4	96.0	96.7	**97.3**	95.2	96.0	93.0	96.8
		aSiMicro	96.2	96.1	95.8	96.2	**97.0**	94.4	95.7	91.8	96.5
		aSiTandem	96.0	96.1	95.8	95.9	**97.0**	93.4	95.5	90.4	96.5
$RMSE$ (°C)	Cocoa	xSi	2.68	3.54	2.73	4.75	**2.29**	3.92	3.21	4.54	2.74
		mSi	2.49	2.86	3.28	3.73	**2.14**	3.08	3.82	3.68	3.50
		CdTe	3.53	4.47	3.06	5.81	3.18	5.09	3.48	5.56	**2.98**
		CIGS	3.50	4.51	2.94	5.86	3.18	5.07	3.29	5.63	**2.77**
		HIT	2.81	3.68	2.84	4.86	**2.41**	3.96	3.30	4.42	2.80
		aSiMicro	2.60	3.31	2.92	4.43	**2.19**	3.67	3.47	4.19	3.04
		aSiTandem	3.95	5.08	3.06	6.49	3.63	5.59	3.27	6.14	**2.68**
	Golden	xSi	3.10	3.49	3.40	4.64	**2.89**	5.39	3.58	6.14	3.39
		mSi	3.23	3.99	2.94	5.43	2.95	6.14	3.12	6.70	**2.80**
		CdTe	3.57	4.32	**3.07**	5.88	3.28	6.87	3.42	7.17	3.08
		CIGS	3.33	3.63	3.63	4.77	**3.01**	5.69	3.90	6.25	3.63
		HIT	3.81	4.55	3.44	5.92	3.49	6.48	3.57	7.07	**3.21**
		aSiMicro	3.58	4.29	3.30	5.66	3.29	6.30	3.45	6.89	**3.12**
		aSiTandem	3.67	4.37	3.37	5.74	3.32	6.42	3.53	6.99	**3.15**
	Eugene	xSi	2.43	2.33	2.88	2.74	**2.15**	5.23	2.66	4.61	2.42
		mSi	2.42	2.35	3.07	2.32	**2.17**	4.57	2.61	3.92	2.40
		CdTe	2.70	2.69	2.94	3.34	**2.41**	5.88	2.87	5.01	2.60
		CIGS	2.73	2.68	3.09	3.13	**2.47**	5.59	2.94	4.84	2.70
		HIT	2.65	2.51	3.30	2.40	**2.40**	4.78	2.88	3.79	2.67
		aSiMicro	2.58	2.50	3.07	2.76	**2.32**	5.21	2.79	4.34	2.56
		aSiTandem	2.77	2.80	2.84	3.66	**2.48**	6.32	2.94	5.46	2.66

References

1. IEA. Solar PV. IEA: Paris, France. Available online: https://www.iea.org/reports/solar-pv (accessed on 24 July 2021).
2. Huld, T.; Gracia Amillo, A.M. Estimating PV module performance over large geographical regions: The role of irradiance, air temperature, wind speed and solar spectrum. *Energies* **2015**, *8*, 5159–5181. [CrossRef]
3. Kratochvil, J.A.; Boyson, W.E.; King, D.L. *Photovoltaic Array Performance Model*; Albuquerque, N.M., Livermore, C.A., Eds.; MDPI: Basel, Switzerland, 2004; Volume 8.
4. Cañete, C.; Carretero, J.; Sidrach-de-Cardona, M. Energy performance of different photovoltaic module technologies under outdoor conditions. *Energy* **2014**, *65*, 295–302. [CrossRef]
5. Kurnik, J.; Jankovec, M.; Brecl, K.; Topic, M. Outdoor testing of PV module temperature and performance under different mounting and operational conditions. *Sol. Energy Mater. Sol. Cells* **2011**, *95*, 373–376. [CrossRef]
6. Gaglia, A.G.; Lykoudis, S.; Argiriou, A.A.; Balaras, C.A.; Dialynas, E. Energy efficiency of PV panels under real outdoor conditions–An experimental assessment in Athens, Greece. *Renew. Energy* **2017**, *101*, 236–243. [CrossRef]

7. Zsiborács, H.; Pintér, G.; Bai, A.; Popp, J.; Gabnai, Z.; Pályi, B.; Farkas, I.; Baranyai, N.H.; Gützer, C.; Trimmel, H.; et al. Comparison of thermal models for ground-mounted south-facing photovoltaic technologies: A practical case study. *Energies* **2018**, *11*, 1114. [CrossRef]
8. Skoplaki, E.; Palyvos, J.A. Operating temperature of photovoltaic modules: A survey of pertinent correlations. *Renew. Energy* **2009**, *34*, 23–29. [CrossRef]
9. Mora Segado, P.; Carretero, J.; Sidrach-de-Cardona, M. Models to predict the operating temperature of different photovoltaic modules in outdoor conditions. *Prog. Photovolt. Res. Appl.* **2015**, *23*, 1267–1282. [CrossRef]
10. Mattei, M.; Notton, G.; Cristofari, C.; Muselli, M.; Poggi, P. Calculation of the polycrystalline PV module temperature using a simple method of energy balance. *Renew. Energy* **2006**, *31*, 553–567. [CrossRef]
11. Akhsassi, M.; El Fathi, A.; Erraissi, N.; Aarich, N.; Bennouna, A.; Raoufi, M.; Outzourhit, A. Experimental investigation and modeling of the thermal behavior of a solar PV module. *Sol. Energy Mater. Sol. Cells* **2018**, *180*, 271–279. [CrossRef]
12. Faiman, D. Assessing the outdoor operating temperature of photovoltaic modules. *Prog. Photovolt. Res. Appl.* **2008**, *16*, 307–315. [CrossRef]
13. Koehl, M.; Heck, M.; Wiesmeier, S.; Wirth, J. Modeling of the nominal operating cell temperature based on outdoor weathering. *Sol. Energy Mater. Sol. Cells* **2011**, *95*, 1638–1646. [CrossRef]
14. Soteris, A. *Kalogirou Solar Energy Engineering: Processes and Systems*; Elsevier Inc.: Amsterdam, The Netherlands, 2009; ISBN 9780123745019.
15. PVsyst 7 Help, Array Thermal Losses. Available online: https://www.pvsyst.com/help/index.html (accessed on 18 March 2021).
16. Skoplaki, E.; Boudouvis, A.G.; Palyvos, J.A. A simple correlation for the operating temperature of photovoltaic modules of arbitrary mounting. *Sol. Energy Mater. Sol. Cells* **2008**, *92*, 1393–1402. [CrossRef]
17. Muller, M.; Marion, B.; Rodriguez, J. Evaluating the IEC 61215 Ed.3 NMOT procedure against the existing NOCT procedure with PV modules in a side-by-side configuration. In Proceedings of the 2012 38th IEEE Photovoltaic Specialists Conference, Austin, TX, USA, 3–8 June 2012; pp. 697–702. [CrossRef]
18. Zouine, M.; Akhsassi, M.; Erraissi, N.; Aarich, N.; Bennouna, A.; Raoufi, M.; Outzourhit, A. Mathematical models calculating PV module temperature using weather data: Experimental study. *Lect. Notes Electr. Eng.* **2019**, *519*, 630–639. [CrossRef]
19. Schwingshackl, C.; Petitta, M.; Wagner, J.E.; Belluardo, G.; Moser, D.; Castelli, M.; Zebisch, M.; Tetzlaff, A. Wind effect on PV module temperature: Analysis of different techniques for an accurate estimation. *Energy Procedia* **2013**, *40*, 77–86. [CrossRef]
20. Vietnam—Solar Radiation Measurement Data obtained from World Bank via ENERGYDATA.info. Available online: https://energydata.info/dataset/vietnam-solar-radiation-measurement-data (accessed on 11 April 2021).
21. Polo, J.; Gastón, M.; Vindel, J.M.; Pagola, I. Spatial variability and clustering of global solar irradiation in Vietnam from sunshine duration measurements. *Renew. Sustain. Energy Rev.* **2015**, *42*, 1326–1334. [CrossRef]
22. Marion, B.; Anderberg, A.; Deline, C.; Del Cueto, J.; Muller, M.; Perrin, G.; Rodriguez, J.; Rummel, S.; Silverman, T.J.; Vignola, F.; et al. New data set for validating PV module performance models. In Proceedings of the 2014 IEEE 40th Photovoltaic Specialist Conference, Denver, CO, USA, 8–13 June 2014; pp. 1362–1366. [CrossRef]
23. Lustbader, J.; Afshin, A. NREL Vehicle Testing and Integration Facility (VTIF): Rotating Shadowband Radiometer (RSR). Golden, Colorado (Data); NREL Report No. DA-5500-56490. Available online: https://midcdmz.nrel.gov/apps/sitehome.pl?site=VTIF (accessed on 26 April 2021).
24. Global Modeling and Assimilation Office (GMAO). MERRA-2 tavg1_2d_slv_Nx: 2d,1-Hourly, Time-Averaged, Single-Level, Assimilation, Single-Level Diagnostics V5.12.4. Goddard Earth Sciences Data and Information Services Center (GES DISC): Greenbelt, MD, USA, 2015. Available online: https://goldsmr4.gesdisc.eosdis.nasa.gov/data/MERRA2/M2T1NXSLV.5.12.4/ (accessed on 27 January 2021).
25. Evans, D.L. Simplified method for predicting photovoltaic array output. *Sol. Energy* **1981**, *27*, 555–560. [CrossRef]
26. Armstrong, S.; Hurley, W.G. A thermal model for photovoltaic panels under varying atmospheric conditions. *Appl. Therm. Eng.* **2010**, *30*, 1488–1495. [CrossRef]
27. Ross, R.G. Interface design considerations for terrestrial solar cell modules. In Proceedings of the 12th Photovoltaic Specialists Conference, Baton Rouge, LA, USA, 15–18 November 1976; pp. 801–806.

applied sciences

MDPI

Article

Hydraulic Vibration and Possible Exciting Sources Analysis in a Hydropower System

Aili Shen [1], Yimin Chen [2], Jianxu Zhou [1,*], Fei Yang [2], Hongliang Sun [2] and Fulin Cai [1]

[1] College of Water Conservancy and Hydropower Engineering, Hohai University, Nanjing 210098, China; ailishenhhu@163.com (A.S.); flcainj@163.com (F.C.)
[2] Huadong Engineering Corporation Limited, Hangzhou 311122, China; chen_ym@hdec.com (Y.C.); yang_f@hdec.com (F.Y.); sun_hl2@hdec.com (H.S.)
* Correspondence: jxzhouhhu@163.com

Abstract: To understand the hydraulic vibration characteristics in a traditional hydropower system and identify possible exciting sources that may induce serious hydraulic vibrations in the flow passage, experimental tests and numerical calculations were conducted for different operating conditions. The experimental results show that the pressure fluctuations are mainly related to the vortex rope phenomena in the draft tube, and the dominant frequency of pressure fluctuation is 0.2~0.4 times the runner rotational frequency (f_n). The numerical results show all the attenuating factors are negative, which indicates the system itself is stable on the condition that all the hydraulic elements have steady operating performance. The free vibration analyses confirm that the frequency range of the vortex rope in the draft tube partly overlaps the natural frequencies of the hydropower system. Apart from the vortex rope, the runner rotational frequency is another common frequency that is approximately equal to the frequency of the 10th vibration mode. From the vibration mode shapes, it is inferred that a small disturbance in its frequency close or equal to a specific natural frequency of the vibration mode could induce large pressure oscillations in the tail tunnel. In light of the system's response to different forcing frequencies, the vortex rope formed under off-design conditions and runner rotational frequency is verified to be the potential exciting source of a traditional hydropower system, and the frequency $0.2 f_n$ is much more dangerous than other disturbances to the system.

Keywords: hydraulic vibration; pressure fluctuation; hydropower system; vortex rope

Citation: Shen, A.; Chen, Y.; Zhou, J.; Yang, F.; Sun, H.; Cai, F. Hydraulic Vibration and Possible Exciting Sources Analysis in a Hydropower System. *Appl. Sci.* **2021**, *11*, 5529. https://doi.org/10.3390/app11125529

Academic Editor: María Isabel Lamas Galdo

Received: 18 May 2021
Accepted: 9 June 2021
Published: 15 June 2021

Publisher's Note: MDPI stays neutral with regard to jurisdictional claims in published maps and institutional affiliations.

Copyright: © 2021 by the authors. Licensee MDPI, Basel, Switzerland. This article is an open access article distributed under the terms and conditions of the Creative Commons Attribution (CC BY) license (https://creativecommons.org/licenses/by/4.0/).

1. Introduction

Nowadays, the increasing global demands for electricity and environmental protections are gradually bringing about the transformation of energy structure such that clean and renewable energy covers an increasing proportion of energy consumption [1]. In this aspect, hydropower plays an irreplaceable role in the clean and renewable energy system, and attracts much attention [2]. Hydraulic turbines are designed for working at the rated head and rated discharge, which is defined as the best efficiency point (BEP), while the current tendency for hydropower plants is that they are undertaking increasing power frequency regulation tasks in the electric grid, and turbines are required to work at off-design conditions more than before [3]. Vibrations exist in various fields, including hydropower stations, and vibration mitigations have been extensively investigated by researchers [4,5]. Recently, more hydraulic vibration phenomena in hydropower systems have been reported accompanied by obvious pressure oscillations along with their hydro-mechanical systems, and even severe accidents, such as local structural damages and power swings, have happened [6].

When turbines operate under off-design conditions, pressure fluctuations due to rotor–stator interaction (RSI), rotating stall, vortex rope, and other flow instabilities might be induced, and accompanying vibration phenomena may be generated [7,8]. Numerous studies, including model experimental tests and numerical simulations, have been conducted to investigate the pressure fluctuations under various off-design conditions. With an overall literature review, it was pointed out that pressure fluctuations in the turbine caused by transient processes would result in fatigue development in the runner, and then shorten the life span of the runner [9]. Based on CFD simulations and validated by experiments, rotor–stator interactions (RSI) were reported to be the primary cause of pressure fluctuation in the vaneless space, and the geometric and operating parameters of the unit were the main affecting factors [10–12]. Moreover, vortex rope in the draft tube was an unavoidable problem that formed under off-design conditions, and this would cause low-frequency but high-amplitude pressure fluctuations in the draft tube that propagate upstream and downstream [13,14]. Qin et al. discovered that the Thoma number influenced not only pressure fluctuation amplitude but also the distribution of the frequency components of pressure fluctuation in the draft tube [15]. Yu et al. discovered that there is a close relation between vortex rope and cavitation in the draft, such tube that the cavitation increases the vortex production as well as the pressure fluctuation frequency induced by vortex rope [16]. Apart from traditional hydraulic turbines, for a pump-turbine operating under off-design conditions, especially in S-shaped regions, pressure fluctuation in the flow passage is also attracting increasing attention [17–19]. Through numerical simulations of a pump-turbine working in an off-design way, it has been reported that the low-frequency pressure fluctuations originate from the rotation of vortices in the draft tube; further, the number of runner blades had an impact on the dominant frequency of pressure fluctuations in the draft tube [20,21]. The vibration affected the security of the hydropower station, and various strategies used to suppress vortex rope oscillation in the turbine flow passage have been explored [22,23]. In view of the pressure fluctuation caused by the vortex rope under off-design conditions, a novel passive control method using an adjustable diaphragm is introduced in decelerating swirling flow; further, the water injection method was proposed to mitigate the pressure fluctuation and change the velocity field in the flow passage, and correlation between the water jet discharge and vortex rope has been investigated [24,25]. However, the experimental method is restricted owing to its high costs and security risk, and the numerical simulation, including the one-dimensional method of characteristics (1D-MOC) and three dimensional (3D) numerical simulations [26] based on computational fluid dynamics (CFD), sometimes requires plenty of computation time, since the process of convergence to a stable state is very slow for long-distance water conveyance systems [27,28].

Hydraulic vibration is a periodic hydraulic transient that exists in various water conveyance systems and may result in instabilities and local destructions [29,30]. With complex condition switches under frequency regulation tasks, both the initial and boundary conditions become extremely complicated, especially for off-design conditions. If frequencies of exciting sources are coincident with the natural frequencies of the hydropower system and continuous working, hydraulic resonance will inevitably develop, and severe pressure oscillations may occur. Consequently, intense flow-induced pressure oscillations may lead to strong vibrations and structural damages [31]. As hydraulic turbines are subjected to increasing off-design operating conditions, the resonance may originate from the vortex rope in the draft tube when its frequency is approximately close to the intrinsic frequency of the water oscillation [32]. However, the exciting sources of hydraulic vibration in hydropower systems have not yet been revealed, and it is imperative to have a clear understanding of the exciting sources of hydraulic vibration in hydropower systems so as to ensure safety and stability during the operating process.

In this paper, pressure fluctuations in the turbine are analyzed, and possible exciting sources that may induce hydraulic vibration in hydropower systems are identified. Firstly, experimental tests were carried out for three off-design conditions by utilizing a reduced scale model runner to monitor pressure fluctuation characteristics in the spiral case, vaneless space, draft tube cone, and draft tube elbow. Secondly, based on transfer matrix and hydraulic impedance methods, free vibration analyses were performed to assess the vibration characteristics, including natural frequencies with attenuating factors of each order, and corresponding vibration mode shapes. Finally, the pressure fluctuations were analyzed in both the time domain and the frequency domain. Then, by comparing the frequency characteristics of pressure fluctuations with natural frequencies of the hydropower system, the correlations between flow instabilities in the hydraulic turbine and hydraulic vibration were revealed, and possible exciting sources were identified.

2. Research Object Description

The research project in this paper involved two turbines with a capacity of 1015 MW per unit. This set up consists of two parallel water diversion penstocks and tail branches, a downstream surge tank, and a D-shaped tail tunnel with a maintenance tail gate shaft located next to the tail water. Figure 1 is a sketch of the plan layout and detailed division information of the hydropower system.

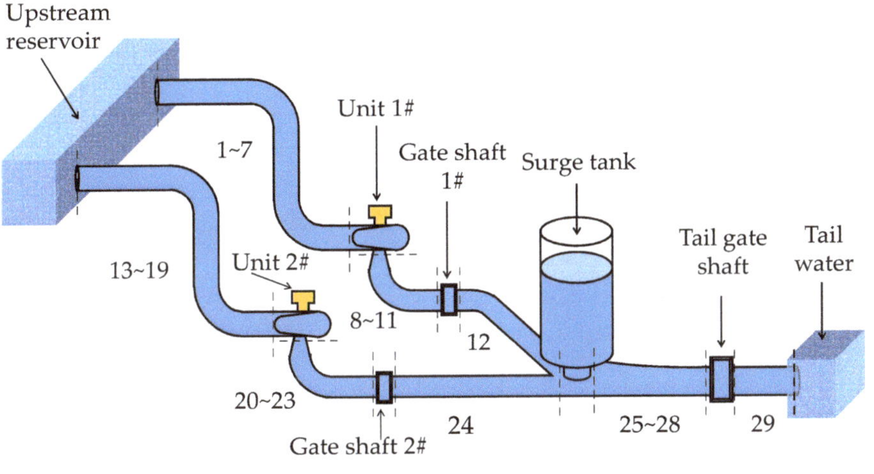

Figure 1. Sketch of the plan layout of the hydropower system.

Table 1 lists the details of each pipe segment, including length and equivalent diameter, in the hydropower system.

Table 1. Basic parameters of the pipe in the hydropower system.

No.	L(m)	D(m)	No.	L(m)	D(m)	No.	L(m)	D(m)
1	41.0	15.925	11	24.55	17.145	21	61.1	11.460
2	166.446	10.999	12	93.25	16.456	22	60.85	17.145
3	47.124	10.203	13	41.0	15.925	23	24.55	17.145
4	104.50	10.203	14	167.938	10.999	24	74.0	16.456
5	47.124	10.203	15	47.124	10.203	25	103.745	17.628
6	31.88	9.053	16	104.50	10.203	26	972.805	17.628
7	42.4	8.600	17	47.124	10.203	27	100.0	17.605
8	20.0	11.460	18	31.88	9.053	28	122.166	17.605
9	61.1	11.460	19	42.4	8.600	29	397.048	17.605
10	60.85	17.145	20	20.0	11.460			

The specific parameters of the prototype turbine are as follows: the rated output P_r = 1015 MW, the rated head H_r = 202 m, the rated discharge Q_r = 545.49 m^3/s, and the rated rotational speed n_r = 111.1 r/min. The maximum and minimum head of the prototype turbine are H_{max} = 243.1 m and H_{min} = 163.9 m, respectively. The test rig consists of a spiral casing, a scaled model turbine runner with 15 blades, and a draft tube.

Figure 2 shows the comprehensive characteristic curves of the turbine. Here, Q_{11} and n_{11} stand for the unit discharge and unit speed, respectively, which are defined as $Q_{11} = Q/(D_1^2\sqrt{H})$ and $n_{11} = nD_1/\sqrt{H}$. From Figure 2, the optimal operating conditions for the turbine are guide vane opening a = 18.8°, unit speed n_{11} = 53.07 r/min, unit flow rate Q_{11} = 0.579 m^3/s, and peak efficiency = 95.07%.

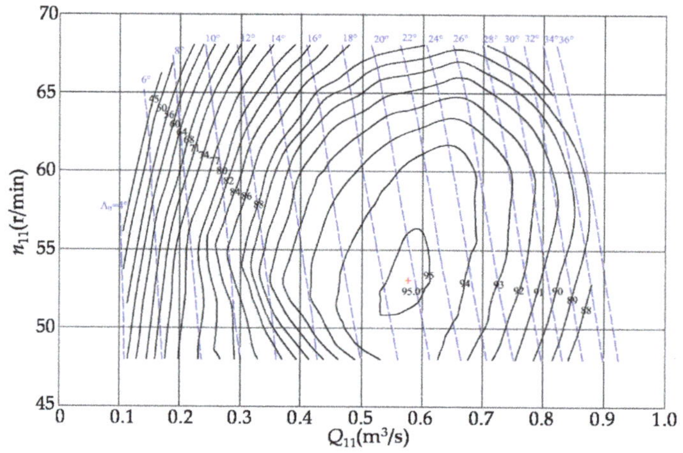

Figure 2. Comprehensive characteristic curves of the turbine.

3. Materials and Mathematical Model

3.1. Governing Equations

For a single pressurized pipe, as shown in Figure 3, the simplified governing equations for the internal steady-oscillatory flow, including the momentum equation and continuity equation, can be deduced by simplifying differential equations of transient flow [33]. The two governing equations are as below,

$$\frac{\partial H(x,t)}{\partial x} + \frac{1}{gA}\frac{\partial Q(x,t)}{\partial t} + \frac{fQ^2(x,t)}{2gDA^2} = 0 \quad (1)$$

$$\frac{\partial Q(x,t)}{\partial x} + \frac{gA}{a^2}\frac{\partial H(x,t)}{\partial t} = 0 \quad (2)$$

where $H = \overline{H} + h$, $Q = \overline{Q} + q$, H is the instantaneous pressure head, Q is the instantaneous discharge, $\overline{H}$ and $\overline{Q}$ are the average value, and h and q are the fluctuation value from the average. x is length to the inlet, t is the time, g is the gravity acceleration, f is the friction factor, A is the cross-sectional area of the pipe, D is the diameter of the pipe, and a is the wave speed.

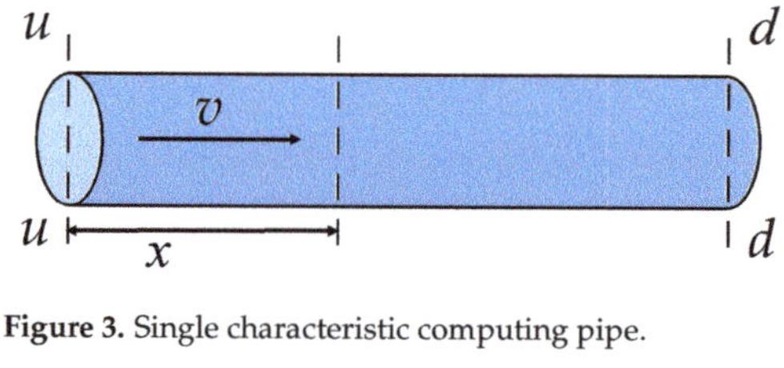

Figure 3. Single characteristic computing pipe.

By substituting the instantaneous items into Equations (1) and (2), and removing the average items, the following form of equations can be obtained and written as

$$\frac{\partial \Phi(x,t)}{\partial x} + B\frac{\partial \Phi(x,t)}{\partial t} + G\Phi(x,t) = 0 \quad (3)$$

where $B = \begin{pmatrix} 0 & L \\ C & 0 \end{pmatrix}$, $G = \begin{pmatrix} 0 & R \\ 0 & 0 \end{pmatrix}$, $\Phi(x,t) = \begin{Bmatrix} h(x,t) \\ q(x,t) \end{Bmatrix}$, $L = 1/(gA)$, $C = Ag/a^2$, $R = f\overline{Q}/(gDA^2)$, L, C, and R are defined as the hydraulic inductance, hydraulic capacitance, and hydraulic resistance of the fluid in the pressurized pipe, respectively.

Application of the Laplace transformation yields the following subsidiary equation,

$$\frac{d\hat{\Phi}(x,s)}{dx} + (Bs+G)\hat{\Phi}(x,s) = 0 \quad (4)$$

where $\hat{\Phi}(x,s) = \{\ \hat{h}(x,s)\ \ \hat{q}(x,s)\ \}^T$ is the transformation of $\Phi(x,t)$, $\hat{\Phi}(x,s) = \int_0^{\infty} \Phi(x,t)e^{-st}dt$. $s = \sigma + i\omega$, s is the complex frequency, σ is coefficient of attenuation, and ω is angular frequency.

Assuming the oscillatory pressure head and discharge at the upstream end are known, the complex oscillatory pressure head and discharge are expressed as

$$\begin{Bmatrix} \hat{h}(x,s) \\ \hat{q}(x,s) \end{Bmatrix} = \begin{pmatrix} \cosh \gamma x & -Z_C \sinh \gamma x \\ -\frac{1}{Z_C}\sinh \gamma x & \cosh \gamma x \end{pmatrix} \begin{Bmatrix} \hat{h}(0,s) \\ \hat{q}(0,s) \end{Bmatrix} \quad (5)$$

where $Z_C = \gamma/(Cs)$ is the characteristic impedance of the pipe, $\gamma = \sqrt{Cs(R+sL)}$ is the complex propagation constant, and x is the length from the inlet.

3.2. *Transfer Matrix and Hydraulic Impedance Methods*

The transfer matrix is a square matrix that relates two state vectors at any point of the pipe by introducing matrix form. Hydraulic impedance is defined as the ratio of the complex head to the complex discharge at the same cross-section. Expressions of some commonly used hydraulic elements have already been established [33,34]. Generally, for various complicated pressurized water conveyance systems, hydraulic vibration analyses are realized by combining the transfer matrix and hydraulic impedance. Two kinds of matrices are commonly used.

3.2.1. Field Matrix

The field transfer matrix connects state vectors at adjacent cross-sections in the same pipe, and as previously deduced, the field matrix of a single pressurized pipe with a length of l is

$$F = \begin{pmatrix} \cosh \gamma l & -Z_C \sinh \gamma l \\ -\sinh \gamma l / Z_C & \cosh \gamma l \end{pmatrix} \quad (6)$$

3.2.2. Point Matrix

The point transfer matrix relates the left and right state vectors of local discontinuity, such as a junction, valve, turbine, etc. The point matrices of several common hydraulic elements are presented below.

For a series junction connecting two pipes, neglecting the minor losses and obeying the relationship $Q_{D1} = Q_{U2}$, $H_{D1} = H_{U2}$, the point matrix is,

$$P_S = \begin{pmatrix} 1 & 0 \\ 0 & 1 \end{pmatrix} \quad (7)$$

The equations for a hydraulic machine operating at a fixed speed can be expressed in a simplified matrix form if there is no excitation pressure head and flow rate at the turbine point,

$$P_T = \begin{pmatrix} 1 & -M \\ 0 & 1 \end{pmatrix} \tag{8}$$

where M is the slope of the head–discharge curve, and the value of M is usually assumed to be a real constant for pumps and turbines if the opening of guide vanes remains constant in the transient regime.

For a throttled surge tank installed in the system shown in Figure 4, the equation of motion is applied to the fluid in the pipeline between the upstream and downstream ends, and the friction term is linearized while the mean flow conditions are subtracted.

$$h_u(t) - h_d(t) - 2kq(t) = \frac{l}{gA_s}\frac{dq(t)}{dt} \tag{9}$$

where k, A_s, and l are the head loss coefficient, cross-sectional area, and water depth in the surge tank, respectively.

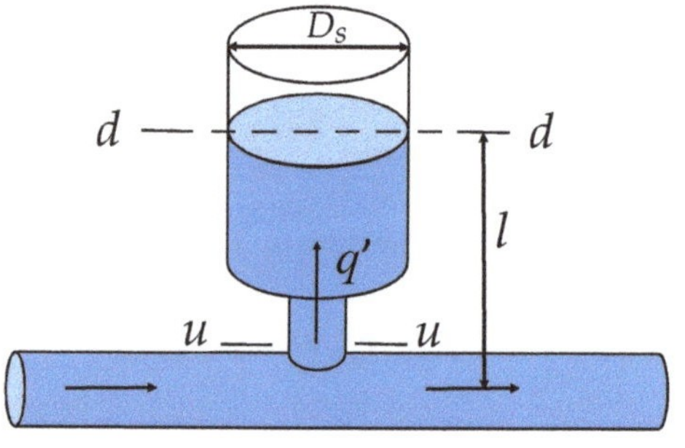

Figure 4. Sketch of the throttled surge tank.

Based on the continuity equation $\frac{dh_d(t)}{dt} = \frac{q}{A_s}$, the Laplace transformation of h_d can be derived, $\hat{h}_d(s) = \frac{\hat{q}(s)}{sA}$.

Applying the Laplace transformation to Equation (9), the impedance formula can be wrtiten,

$$Z_u = \frac{\hat{h}_u(s)}{\hat{q}(s)} = \frac{1}{sA_s} + 2k + \frac{sl}{gA_s} \tag{10}$$

Therefore, the matrix of a throttled surge tank is

$$P_{st} = \begin{pmatrix} 1 & 0 \\ -\frac{1}{Z_u} & 1 \end{pmatrix} \tag{11}$$

For the parallel system in Figure 5 with no forcing function, the field matrix of path j is expressed as $[F]_j = \begin{bmatrix} e_{11} & e_{12} \\ e_{21} & e_{22} \end{bmatrix}_j$. Then, the overall field transfer matrix of the whole parallel system is

$$[F]_{\text{loop}} = \begin{bmatrix} \frac{\zeta}{\eta} & \frac{1}{\eta} \\ \frac{\zeta\xi}{\eta} - \eta & \frac{\xi}{\eta} \end{bmatrix} \tag{12}$$

where η, ζ and ξ are $\eta = \sum_{j=1}^{n} \left(\frac{1}{e_{12}}\right)_j$, $\zeta = \sum_{j=1}^{n} \left(\frac{e_{11}}{e_{12}}\right)_j$, and $\xi = \sum_{j=1}^{n} \left(\frac{e_{22}}{e_{12}}\right)_j$, respectively.

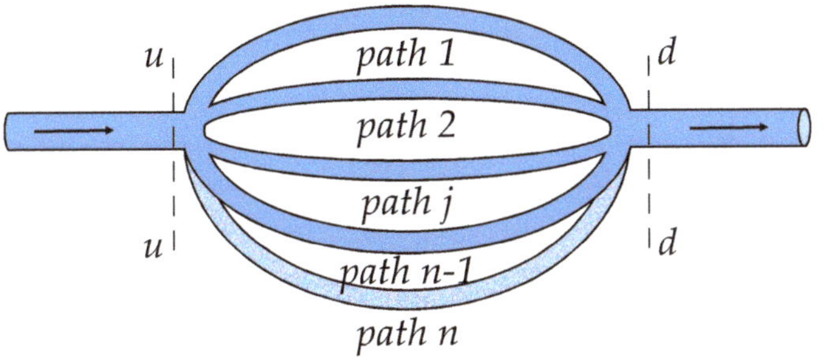

Figure 5. Parallel system.

3.3. Hydraulic Vibration Analysis

3.3.1. Free Vibration Analysis

Free vibration is the residual oscillation in the absence of an external disturbance, and the aim of conducting free vibration analysis is to obtain the complex frequencies of a given system, as well as the corresponding vibration mode shapes. With some commonly used hydraulic elements, the characteristic equations of the whole pressurized water conveyance system can be derived, and the overall matrix that relates the boundary data at two terminal points of the system is expressed as

$$\left\{ \begin{array}{c} \hat{h}(l,s) \\ \hat{q}(l,s) \end{array} \right\} = \left[\begin{array}{cc} u_{11} & u_{12} \\ u_{21} & u_{22} \end{array} \right] \left\{ \begin{array}{c} \hat{h}(0,s) \\ \hat{q}(0,s) \end{array} \right\} \tag{13}$$

For the condition wherein the upstream is a reservoir and the outlet is a tailwater, and water elevation remains constant, $H_U = 0$, $H_D = 0$, which means the impedance values at the inlet and outlet are both zero. Hence, the characteristic equation of the system is written as below,

$$\left\{ \begin{array}{c} Z_D(s) = u_{12}/u_{22} = 0 \\ u_{12} = 0 \end{array} \right. \tag{14}$$

By solving Equation (14), we can obtain the complex frequency s_k (k = 1,2,3, …).

3.3.2. Forced Vibration Analysis

A periodic external disturbance that continues acting on a certain boundary, and which can generate steady oscillatory flow, is recognized as forced vibration. The purpose of forced vibration analysis is to investigate the system's response to a known forcing function that existed all the time. In a fully developed forced vibration, the whole system oscillates with the frequency of the forcing function, and the complex frequency only contains an imaginary part, $s = i\omega_k$, the attenuate factor, while $\sigma = 0$ means the vibration amplitude was independent of time.

3.4. Mathematical Model of Pressurized Flow in the System

For the hydropower system shown in Figure 1, based on the transfer matrix and hydraulic impedance method, the overall transfer matrix of the entire system was built. The format of the overall transfer matrix connecting the oscillatory pressure head and discharge at the upstream and downstream end of the system is,

$$U = \left[\begin{array}{cc} u_{11} & u_{12} \\ u_{21} & u_{22} \end{array} \right] = [F]_{29}[P]_{wz}[F]_{28}[F]_{27}[F]_{26}[F]_{25}[P]_{st}[F]_{loop} \tag{15}$$

where $[F]_i (i = 25, 26, 27, 28, 29)$ is the field matrix of the pipe numbered i in the main tail tunnel, $[P]_{wz}$ is the point matrix of the gate shaft, $[P]_{st}$ is the point matrix of the surge tank located at the branch, and $[F]_{loop}$ is the field matrix of two parallel branches. The corresponding hydraulic vibration analysis process is presented in Figure 6.

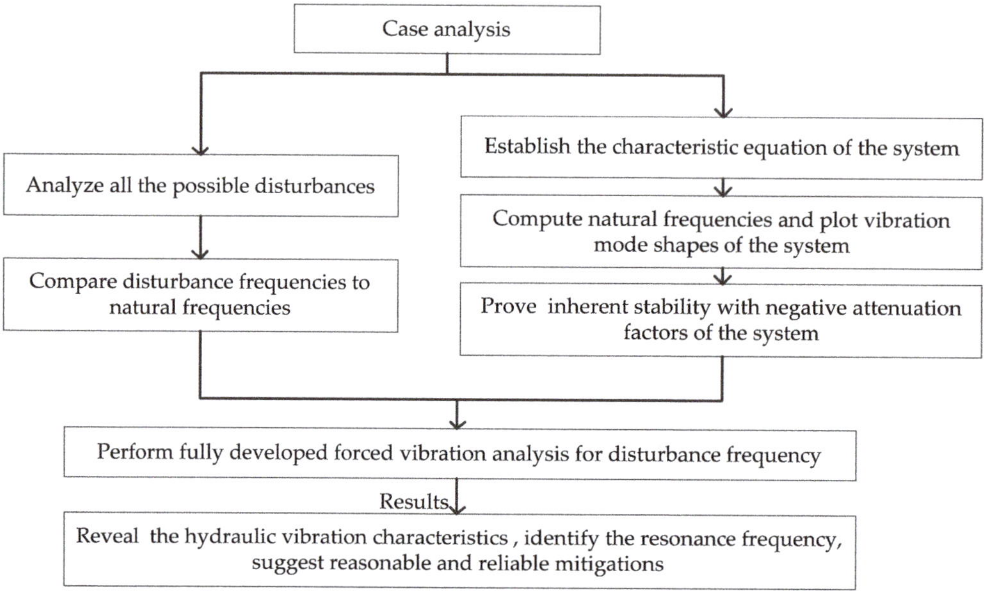

Figure 6. Hydraulic vibration analysis method.

4. Results and Discussions

In the experiment, twelve pressure-monitoring points, designated as CH0~CH12, were evenly arranged in the flow passage as shown in Figure 7. In detail, CH0 and CH1 were located in the spiral casing, CH2~CH5 were located in the vaneless space, CH6~CH9 were located in the draft tube cone, and CH10~CH11 were located in the draft tube elbow.

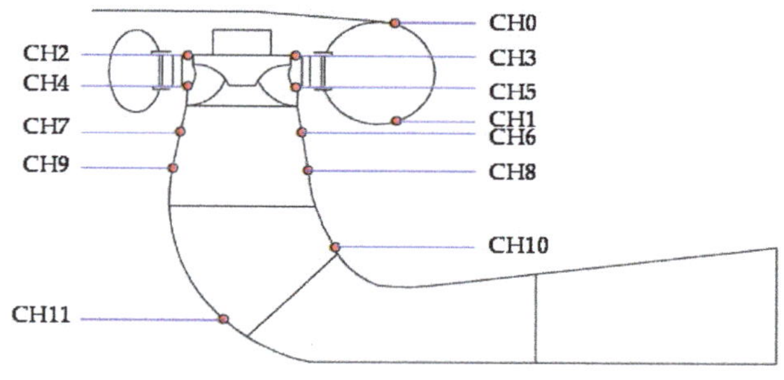

Figure 7. Locations of pressure-monitoring points.

The fluctuation peak $\Delta H/H$ was calculated by processing the pressure signal at the confidence level of 97%, and its definition is

$$\frac{\Delta H}{H} = \frac{p_i - \overline{p}}{1000\rho g H} \times 100\% \quad (16)$$

where H is the rated head of the turbine, p_i is the instantaneous pressure, and $\overline{p}$ is the average pressure.

Three off-design operating conditions are selected in this paper to study the regularity of the pressure fluctuation. Details of operating points for the model runner are listed in Table 2.

Table 2. Details of operating conditions.

Mode	a (°)	n_{11} (r/min)	Q_{11} (L/s)	H_m (m)	P (MW)	P_r (%Pt)
S01	7.88	60.21	145.99	29.91	301.11	29.67
S02	12.63	60.15	249.77	30.32	607.72	59.87
S03	14.12	73.16	260.64	29.88	346.43	48.32

4.1. Pressure Fluctuation in the Flow Passage

The pressure fluctuations at points CH0~CH11 under condition S01, S02, and S03 were monitored and recorded. Table 3 lists the pressure fluctuation peaks in the time domain at points CH0~CH11, and it is clear that the pressure fluctuations in the spiral case are relatively small compared to other points. The corresponding dominant frequencies are listed in Table 4, and $0.2f_n$ is the frequency with the greatest number of occurrences in the vaneless space and draft tube. Consequently, the pressure fluctuations in the vaneless space and draft tube are analyzed in the following section.

Table 3. Pressure fluctuation peaks in time domain under different operating conditions.

Mode	CH0	CH1	CH2	CH3	CH4	CH5	CH6	CH7	CH8	CH9	CH10	CH11
S01	1.88	1.86	1.84	1.78	1.51	1.74	3.95	3.53	3.12	3.13	1.81	2.86
S02	1.91	1.88	2.67	3.02	2.14	2.54	2.68	2.86	2.63	3.36	1.54	2.28
S03	1.27	1.25	2.60	2.50	1.71	2.38	4.34	3.69	3.78	4.33	2.28	4.14

Table 4. f/f_n in frequency domain at different monitoring points.

Mode	CH0	CH1	CH2	CH3	CH4	CH5	CH6	CH7	CH8	CH9	CH10	CH11
S01	0.1	0.1	15	15	0.2	0.2	1.6	1.5	0.2	0.2	0.2	0.2
S02	0.0	0.1	0.2	0.2	1.0	0.2	0.2	0.2	0.4	0.4	0.2	0.2
S03	0.0	0.0	0.2	0.2	0.2	0.2	0.2	0.2	0.2	0.2	0.2	0.2

4.1.1. Pressure Fluctuation in Vaneless Space

Figure 8 shows the pressure fluctuation characteristics in the vaneless space (CH2 and CH4). The time domain characteristics show that the fluctuation peak monitored under mode S02 was the largest compared to S01 and S03. The frequency domain characteristics indicate that under mode S01 with smaller guide vane openings, a blade frequency of $15f_n$ was manifested, caused by rotor–stator interaction, and with guide vane openings increasing, the runner rotational frequency reached $1.0f_n$ at point CH4 under S02. A frequency of $0.2f_n$, affected by vortex rope in the draft tube, persists in all operating modes.

4.1.2. Pressure Fluctuation in the Draft Tube

When turbines work under off-design conditions, the rotational velocity component of fluid in the runner outlet will probably cause vortex rope accompanied by low-frequency and large-amplitude pressure fluctuations in the draft tube. According to the empirical formula, the estimated frequency range of vortex rope is $f = (0.167\text{~}0.5)f_n$, and the frequency of $0.2\text{~}0.4f_n$ captured in the experiment is in the range of the estimated frequency. Figure 9 shows the pressure fluctuations in the draft tube (CH6) under S01, S02, and S03. The time domain characteristics illustrate that the pressure fluctuation in the draft tube cone is the highest compared to other regions. From the frequency domain characteristics, we can infer that $0.2f_n$ is the leading frequency that occurred in all conditions. Additionally, frequencies of $1.6f_n$ and $1.5f_n$, measured at the inlet of the draft tube only, occurred in S01, and these may propagate from the upstream region affected by RSI.

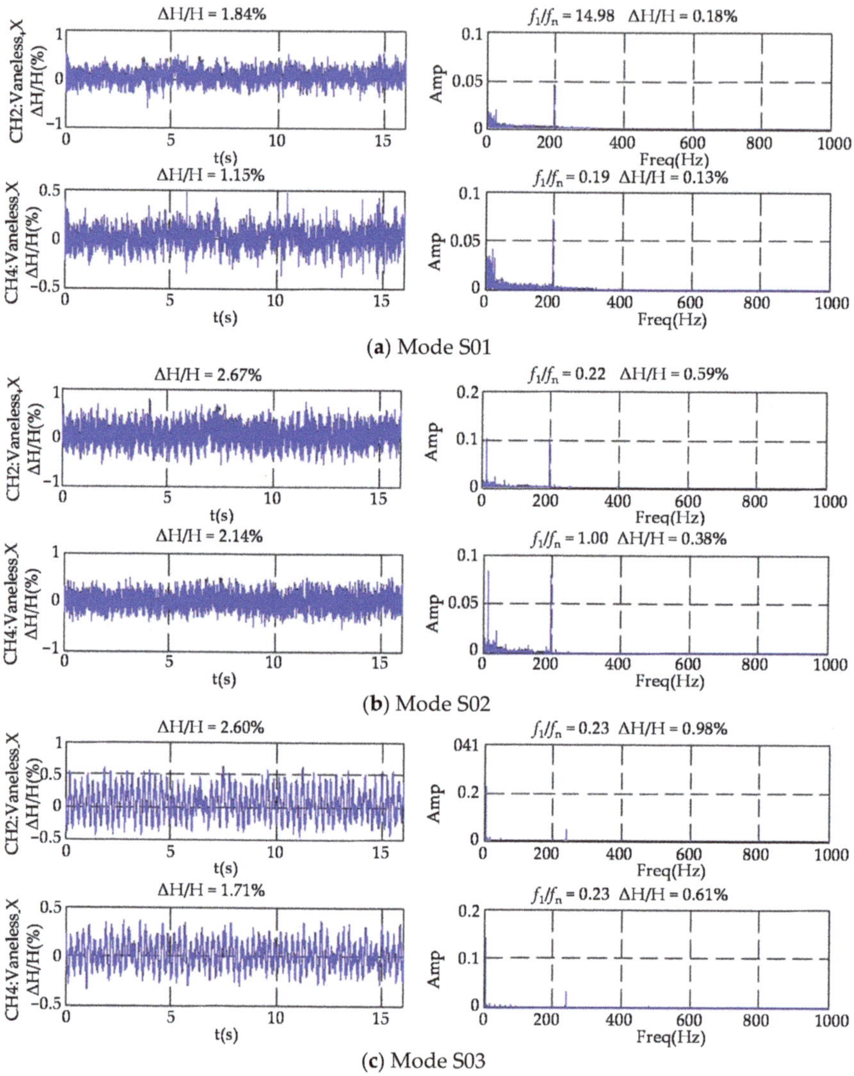

Figure 8. Pressure fluctuations in vaneless space.

4.2. Numerical Computation

4.2.1. Natural Frequencies

On the derivation of the mathematical model used for hydraulic vibration analysis, as mentioned above, further free vibration analyses were performed. The complex frequency mainly depends on the length of the system and the wave speed; as a result, for a traditional pressurized hydropower system, the hydraulic vibration characteristics can be revealed at a typical designed operating point. For this research project, the boundary conditions were H_u = 806.8 m and H_d = 597.42 m, and both turbines worked at the rated head and discharge. Then, by solving the characteristic Equation (14), the complex frequencies were calculated and are listed in the Case 1 column in Table 5. Taking the effect of wave speed on numerical computation into consideration, the column Case 2 lists the results by adding a 10% increase to the wave speed.

The data in Table 5 show that the frequencies of the first three orders have few relationships to wave speed, whereas for orders higher than 3, the frequencies are affected by the wave speed distinctly, and with increasing orders, the deviation between two frequencies corresponding to the same order becomes obvious. For the frequency 7.4851 of 8 order in Case 1 is approximately equal to the frequency of 7 order in Case 2 if ignoring the difference after the decimal point, so it is unnecessary to perform free vibration analysis on higher orders because of the clearly error induced by value of wave speed, and herein, only frequencies lower than f_n should be reserved in Table 5. Further, all attenuating factors in Table 5 are negative, which means that all vibration modes would attenuate with time until a steady state is achieved, and the possibility of self-excited resonance can be excluded.

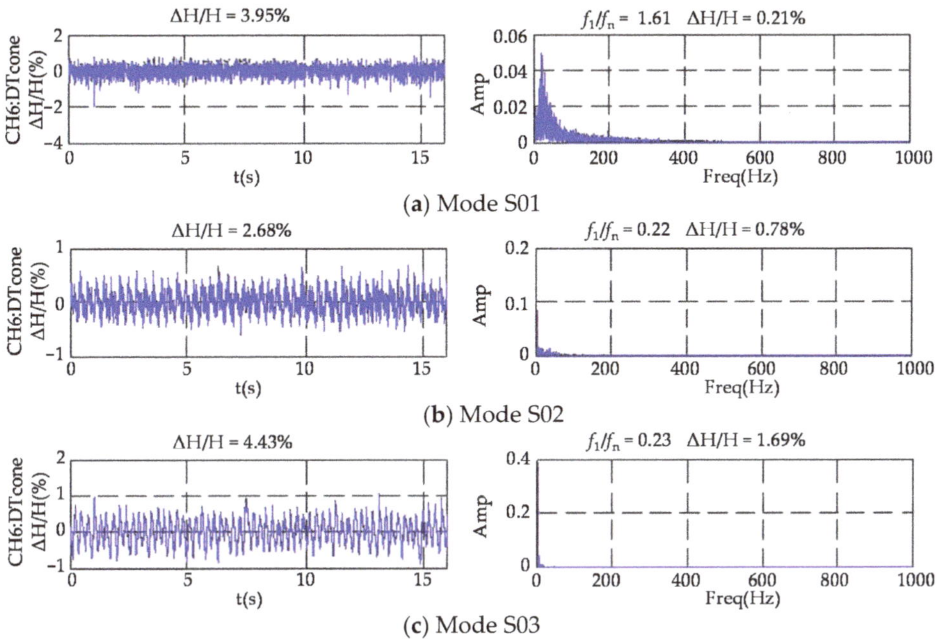

Figure 9. Pressure fluctuations in the draft tube.

Table 5. Complex frequencies of the hydropower system.

Order *k*th	Case 1		Case 2	
	σ	ω	σ	ω
1	−0.0046	0.0284	−0.0046	0.0284
2	−0.0048	0.2597	−0.0048	0.2599
3	−0.0025	0.3640	−0.0025	0.3640
4	−0.0038	2.3428	−0.0038	2.6017
5	−0.0036	4.6627	−0.0036	5.1799
6	−0.3048	5.5526	−0.3041	6.1735
7	−0.0044	6.8170	−0.0044	7.5732
8	−0.0055	7.4851	−0.0055	8.3165
9	−0.0265	9.4424	−0.0276	10.4924
10	−0.0034	11.6914	−0.0034	12.9902

4.2.2. Mode Shapes

The vibration mode shapes are calculated with an assumed initial oscillatory discharge value Q_U = 1.0 m^3/s at the upstream end. The rotational frequency of the prototype runner is f_n = 1.852 Hz, and ω = 11.63 rad/s. According to the empirical formula, the estimated frequency range of vortex rope was f = (0.167~0.5) f_n = 0.309~0.926 Hz, and ω = 1.94~5.82 rad/s. Based on the above analysis, only orders whose angular frequency is in the range of vortex rope frequency (4th, 5th, 6th) and close to the runner rotational frequency (10th) are selected for plotting in Figure 10, including oscillatory discharge and oscillatory pressure head, and the abscissa is the distance from the inlet, while the black line is the head and the red line is the flow rate. For frequencies approximately equal to those of the 4th, 5th, and 10th orders, a small disturbance could cause intense pressure oscillation in the tail tunnel. However, for the sixth order, there was no obvious pressure oscillation in the tail tunnel shown in Figure 10c. The mode shapes reveal the modulus of oscillatory value along the pipeline at different frequencies, and the amplitude of pressure fluctuation is at a minimum at the node and a maximum at the antinode. The locations of the node and antinode of the flow are opposite to those in the head. Since the upstream and downstream are reservoirs with a constant water level, the oscillatory head is zero at both ends and in all vibration modes.

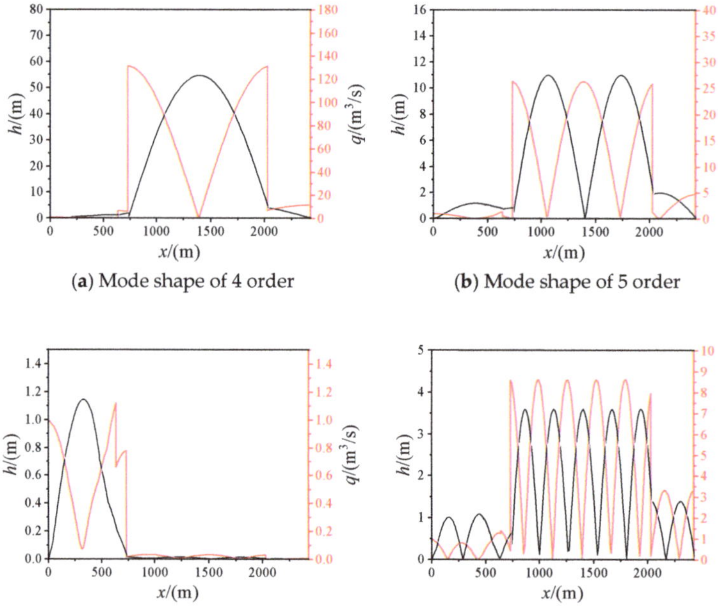

(**a**) Mode shape of 4 order (**b**) Mode shape of 5 order

(**c**) Mode shape of 6 order (**d**) Mode shape of 10 order

Figure 10. Vibration mode shapes.

4.3. Comparative Analysis

The estimated frequency range of the vortex rope overlaps the 4th, 5th, and 6th orders, and the rotational frequency of the prototype runner is close to the frequency of the 10th mode. For the prototype turbine, the frequencies of 0.2~0.4f_n measured for the vortex rope and the runner rotational frequency are emphasized here. Forced vibration was performed at three frequencies (0.2f_n, 0.4f_n, and f_n) of oscillating discharge at the turbine point as the forcing function, with an expression $q' = 0.2 \sin \omega t$, and the system's responses are shown in Figure 11.

It is shown in Figure 11 that if the frequency of disturbance close is to the 4th, 5th, or 10th mode, a small disturbance would cause intense pressure oscillation in the tail tunnel, and the response to 0.2f_n is the highest. On the contrary, the response to the runner rotational frequency is the lowest. It is clear that the oscillatory crest value of both pressure and discharge decreases as the disturbance's frequency increases, which indicates that the lower frequency vibration is more severe and should be avoided, because severe pressure oscillation can burst or collapse the pipe due to pressure in excess of the designed pressure. The pressure fluctuation in the tail tunnel shown in Figure 11 is much higher than that in the upstream part of the system. For the blade frequency of $15f_n$ with an extremely short exciting period, it is unnecessary to carry out targeted analyses, since the natural frequency of the higher order is not exact owing to the error in the estimated wave speed, and the corresponding higher-order vibration usually manifests as energy dissipation. It is confirmed that the leading frequencies of the vortex rope and the runner rotational frequency are closely related to natural frequencies, which may induce huge pressure fluctuations and even resonance along the water conveyance line. Under the actual operating conditions, real-time monitoring should concentrate on the frequency characteristics of vortex rope in the draft tube and pressure fluctuation in the vaneless space, in case these equal the natural frequency of the system.

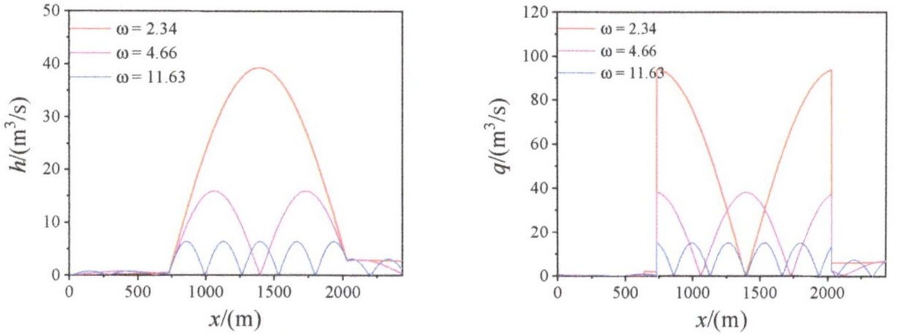

(**a**) Pressure oscillation (**b**) Discharge oscillation

Figure 11. System's response to the forcing function.

5. Conclusions

In this paper, the pressure fluctuations in a reduced-scale model turbine test rig under three sets of off-design conditions are tested. According to the time domain and frequency domain analysis, the leading frequencies of pressure pulsation throughout the flow passage are obtained. Then, the natural frequencies are calculated, and the vibration mode shapes corresponding to various frequencies are revealed for the whole hydraulic system, based on detailed free vibration analysis. It was found that the leading frequencies of the vortex rope in the draft tube partially overlap the natural frequencies, including the fourth, fifth, and sixth modes. It is concluded from the forced vibration analysis that if the frequency of the vortex rope is close to these modes and continues acting as a forcing disturbance, intense pressure fluctuation inevitably occurs, and a frequency of $0.2f_n$ in the vortex rope is the most dangerous disturbance, as this will cause huge pressure fluctuations in the water conveyance line. Besides this, the runner rotational frequency cannot be ignored either, as this may cause severe pressure fluctuations in the tail tunnel. According to the computation results, when the disturbance frequencies are similar to certain natural frequencies, vibration mitigations actions should be taken during the operating stage.

Hydraulic vibration analysis can provide a reference to recognize disturbances during the design stage in order to avoid severe pressure fluctuation, and even resonance, during the operation stage. A future study will focus on the exploration of vibration reduction methods, and on exactly identifying the disturbance by decomposing the pressure signal in the flow passage into synchronous and asynchronous parts. The study will also focus on the wide application of hydraulic vibration theory to various kinds of water conveyance systems.

Author Contributions: Conceptualization, A.S. and J.Z.; methodology, A.S. and J.Z.; software, A.S. and J.Z.; validation, J.Z.; formal analysis, A.S. and J.Z.; investigation, A.S. and J.Z.; resources, Y.C. and F.C.; data curation, J.Z.; writing—original draft preparation, A.S. and J.Z.; writing—review and editing, A.S. and J.Z.; visualization, Y.C.; supervision, J.Z.; project administration, Y.C. and F.Y. and H.S.; funding acquisition, Y.C. and F.Y. and H.S. All authors have read and agreed to the published version of the manuscript.

Funding: This research received no external funding.

Acknowledgments: The authors wish to thank the support given by College of Water Conservancy and Hydropower Engineering, Hohai University and Huadong Engineering Corporation Limited.

Conflicts of Interest: The authors declare no conflict of interest.

References

1. Li, X.; Chen, Z.; Fan, X.; Cheng, Z. Hydropower development situation and prospects in China. *Renew. Sustain. Energy Rev.* **2018**, *82*, 232–239. [CrossRef]
2. Bilgili, M.; Bilirgen, H.; Ozbek, A.; Ekinci, F.; Demirdelen, T. The role of hydropower installations for sustainable energy development in Turkey and the world. *Renew. Energy* **2018**, *126*, 755–764. [CrossRef]
3. Trivedi, C.; Gogstad, P.J.; Dahlhaug, O.G. Investigation of the unsteady pressure pulsations in the prototype Francis turbines—Part 1: Steady state operating conditions. *Mech. Syst. Signal Process.* **2018**, *108*, 188–202. [CrossRef]
4. Glowacz, A.; Glowacz, W.; Kozik, J.; Piech, K.; Gutten, M.; Caesarendra, W.; Liu, H.; Brumercik, F.; Irfan, M.; Faizal Khan, Z. Detection of Deterioration of Three-phase Induction Motor using Vibration Signals. *Meas. Sci. Rev.* **2019**, *19*, 241–249. [CrossRef]
5. Li, D.; Fu, X.; Zuo, Z.; Wang, H.; Li, Z.; Liu, S.; Wei, X. Investigation methods for analysis of transient phenomena concerning design and operation of hydraulic-machine systems—A review. *Renew. Sustain. Energy Rev.* **2019**, *101*, 26–46. [CrossRef]
6. Müller, A.; Favrel, A.; Landry, C.; Yamamoto, K.; Avellan, F. On the physical mechanisms governing self-excited pressure surge in Francis turbines. *IOP Conf. Ser. Earth Environ. Sci.* **2014**, *22*. [CrossRef]
7. Trivedi, C. Time-dependent inception of vortex rings in a Francis turbine during load variation: Large eddy simulation and experimental validation. *J. Hydraul. Res.* **2019**, *58*, 790–806. [CrossRef]
8. Zhang, L.; Wu, Q.; Ma, Z.; Wang, X. Transient vibration analysis of unit-plant structure for hydropower station in sudden load increasing process. *Mech. Syst. Signal Process.* **2019**, *120*, 486–504. [CrossRef]
9. Trivedi, C.; Gandhi, B.; Michel, C.J. Effect of transients on Francis turbine runner life: A review. *J. Hydraul. Res.* **2013**, *51*, 121–132. [CrossRef]
10. Goyal, R.; Gandhi, B.K.; Cervantes, M.J. Transient Pressure Measurements in the Vaneless Space of a Francis Turbine during Load Acceptances from Minimum Load. *J. Phys. Conf. Ser.* **2018**, *1042*. [CrossRef]
11. Guo, L.; Liu, J.; Wang, L.; Qin, D.; Wei, X. Pressure fluctuation propagation of a pump turbine at pump mode under low head condition. *Sci. China Technol. Sci.* **2014**, *57*, 811–818. [CrossRef]
12. Zuo, Z.; Liu, S.; Sun, Y.; Wu, Y. Pressure fluctuations in the vaneless space of High-head pump-turbines—A review. *Renew. Sustain. Energy Rev.* **2015**, *41*, 965–974. [CrossRef]
13. Cheng, H.; Zhou, L.; Liang, Q.; Guan, Z.; Liu, D.; Wang, Z.; Kang, W. A method of evaluating the vortex rope strength in draft tube of Francis turbine. *Renew. Energy* **2020**, *152*, 770–780. [CrossRef]
14. Kumar, S.; Cervantes, M.J.; Gandhi, B.K. Rotating vortex rope formation and mitigation in draft tube of hydro turbines—A review from experimental perspective. *Renew. Sustain. Energy Rev.* **2021**, *136*. [CrossRef]
15. Qin, D.; Xu, Y.; Liu, W.; Wei, X.; Zhao, Y.; Meng, X. Experimental study of the influence of Thoma number and model testing head on pressure fluctuation in draft tube of a Francis turbine. *IOP Conf. Ser. Earth Environ. Sci.* **2016**, *49*. [CrossRef]
16. Yu, A.; Zou, Z.; Zhou, D.; Zheng, Y.; Luo, X. Investigation of the correlation mechanism between cavitation rope behavior and pressure fluctuations in a hydraulic turbine. *Renew. Energy* **2020**, *147*, 1199–1208. [CrossRef]
17. Zhang, X.; Zeng, W.; Cheng, Y.; Yang, Z.; Chen, Q.; Yang, J. Mechanism of Fast Transition of Pressure Pulsations in the Vaneless Space of a Model Pump-Turbine During Runaway. *J. Fluid Eng.* **2019**, *141*. [CrossRef]
18. Zhang, W.; Chen, Z.; Zhu, B.; Zhang, F. Pressure fluctuation and flow instability in S-shaped region of a reversible pump-turbine. *Renew. Energy* **2020**, *154*, 826–840. [CrossRef]
19. Zhang, X.; Cheng, Y.; Yang, Z.; Chen, Q.; Liu, D. Influence of rotational inertia on the runner radial forces of a model pump-turbine running away through the S-shaped characteristic region. *IET Renew. Power Gener.* **2020**, *14*, 1883–1893. [CrossRef]
20. Li, D.; Sun, Y.; Zuo, Z.; Liu, S.; Wang, H.; Li, Z. Analysis of Pressure Fluctuations in a Prototype Pump-Turbine with Different Numbers of Runner Blades in Turbine Mode. *Energies* **2018**, *11*, 1474. [CrossRef]
21. Li, D.; Wang, H.; Qin, Y.; Li, Z.; Wei, X.; Qin, D. Mechanism of high amplitude low frequency fluctuations in a pump-turbine in pump mode. *Renew. Energy* **2018**, *126*, 668–680. [CrossRef]
22. Su, W.-T.; Binama, M.; Li, Y.; Zhao, Y. Study on the method of reducing the pressure fluctuation of hydraulic turbine by optimizing the draft tube pressure distribution. *Renew. Energy* **2020**, *162*, 550–560. [CrossRef]
23. Zhu, L.; Zhang, R.-Z.; Yu, A.; Lu, L.; Luo, X.-W. Suppression of vortex rope oscillation and pressure vibrations in Francis turbine draft tube using various strategies. *J. Hydrodyn.* **2021**. [CrossRef]
24. Bosioc, A.I.; Tănasă, C. Experimental study of swirling flow from conical diffusers using the water jet control method. *Renew. Energy* **2020**, *152*, 385–398. [CrossRef]

25. Tănasă, C.; Bosioc, A.; Muntean, S.; Susan-Resiga, R. A Novel Passive Method to Control the Swirling Flow with Vortex Rope from the Conical Diffuser of Hydraulic Turbines with Fixed Blades. *Appl. Sci.* **2019**, *9*, 4910. [CrossRef]
26. Kan, K.; Zheng, Y.; Chen, H.; Zhou, D.; Dai, J.; Binama, M.; Yu, A. Numerical simulation of transient flow in a shaft extension tubular pump unit during runaway process caused by power failure. *Renew. Energy* **2020**, *154*, 1153–1164. [CrossRef]
27. Yin, C.; Yang, J.; Zeng, W.; Cheng, Y. Simulation of the simultaneous load rejection processes of two parallel pump turbines using a 1D–3D approach. *IOP Conf. Ser. Earth Environ. Sci.* **2019**, *240*. [CrossRef]
28. Zhang, X.; Cheng, Y.; Yang, Z.; Chen, Q.; Liu, D. Water column separation in pump-turbine after load rejection: 1D-3D coupled simulation of a model pumped-storage system. *Renew. Energy* **2021**, *163*, 685–697. [CrossRef]
29. Palikhe, S.; Zhou, J.; Bhattarai, K.P. Hydraulic Oscillation and Instability of a Hydraulic System with Two Different Pump-Turbines in Turbine Operation. *Water* **2019**, *11*, 692. [CrossRef]
30. Zhou, J.; Chen, Y. Discussion on Stochastic Analysis of Hydraulic Vibration in Pressurized Water Diversion and Hydropower Systems. *Water* **2018**, *10*, 353. [CrossRef]
31. Landry, C.; Favrel, A.; Müller, A.; Nicolet, C.; Avellan, F. Local wave speed and bulk flow viscosity in Francis turbines at part load operation. *J. Hydraul. Res.* **2016**, *54*, 185–196. [CrossRef]
32. Wu, Y.; Liu, S.; Dou, H.-S.; Wu, S.; Chen, T. Numerical prediction and similarity study of pressure fluctuation in a prototype Kaplan turbine and the model turbine. *Comput. Fluids* **2012**, *56*, 128–142. [CrossRef]
33. Wiley, E.B.; Streeter, V.L.; Suo, L. *Fluid Transients in Systems*; Prentice Hall: Englewood Cliffs, NJ, USA, 1993.
34. Chaudhry, M.H. *Applied Hydraulic Transients*; Springer: Berlin, Germany, 2014.

applied sciences

MDPI

Article

Computational Modeling of the Thermal Behavior of a Greenhouse

Bruno Lebre [1], Pedro D. Silva [1,2], Luís C. Pires [1,2] and Pedro D. Gaspar [1,2,*]

[1] Department of Electromechanical Engineering, University of Beira Interior, 6201-001 Covilhã, Portugal; bruno.lebre@ubi.pt (B.L.); dinho@ubi.pt (P.D.S.); pires@ubi.pt (L.C.P.)
[2] C-MAST-Center for Mechanical and Aerospace Science and Technologies, 6201-001 Covilhã, Portugal
* Correspondence: dinis@ubi.pt

Abstract: The need for production of all kinds of crops in high quantities and over the entire year makes the agricultural sector one of the major energy consumers. The optimization of this consumption is essential to guarantee its sustainability. The implementation of greenhouses is a strategy that allows assurance of production needs and possesses large optimization potential for the process. This article studies different greenhouse structures by computational simulation using EnergyPlus and DesignBuilder. First, a comparison was performed between the computational results and the measured values from a greenhouse prototype at different operating conditions. Overall, the comparison shows that the computational tool can provide a reasonable prediction of the greenhouse thermal behavior, depending on the differences between the weather data modeled and observed. An outdoor air temperature difference of 16 °C can cause a difference of about 10 °C between the air temperature predicted and measured inside the greenhouse. Subsequently, a selected set of case studies was developed in order to quantify their influence on the thermal performance of the greenhouse, namely: the greenhouse configuration and orientation; the variation of indoor air renewal; changes to the characteristics of the roof; the effect of the thermal mass of the walls; and location of the greenhouse. The results show that a correct greenhouse orientation, together with a polyethylene double cover with a 13 mm air layer, a granite wall of 40 cm thickness on the north wall, and variable airflow rate, may lead to a reduction of the greenhouse energy consumption by 57%, if the greenhouse is located in Lisbon, or by 43%, if it is located in Ostersund, during the harshest months of the heating season.

Keywords: greenhouse thermal performance; numerical study; energy efficiency

Citation: Lebre, B.; Silva, P.D.; Pires, L.C.; Gaspar, P.D. Computational Modeling of the Thermal Behavior of a Greenhouse. *Appl. Sci.* **2021**, *11*, 11816. https://doi.org/10.3390/app112411816

Academic Editors: Hassane Naji, María Isabel Lamas Galdo and Rodriguez J.D.

Received: 29 October 2021
Accepted: 10 December 2021
Published: 13 December 2021

Publisher's Note: MDPI stays neutral with regard to jurisdictional claims in published maps and institutional affiliations.

Copyright: © 2021 by the authors. Licensee MDPI, Basel, Switzerland. This article is an open access article distributed under the terms and conditions of the Creative Commons Attribution (CC BY) license (https://creativecommons.org/licenses/by/4.0/).

1. Introduction

A greenhouse allows the creation of a controlled environment with proper microclimate conditions required for crop growth, increasing crop production rates and quality [1]. Parameters such as indoor air temperature, soil temperature, relative humidity, light intensity, and carbon dioxide concentration can be controlled using a greenhouse. As greenhouses allow cultivation in areas where the natural conditions are unfavorable for plant growth, they may avoid the need to transport vegetables and horticultural crops from distant places [2]. Greenhouses are also being used worldwide for drying, with many advantages concerning the quality of the dry products when compared to traditional drying methods. In this specific case, products are spread on the ground, exposing them directly to solar radiation, with considerable losses due to dirt, dust, insects, microorganisms, animals, and birds [3].

Parameters such as greenhouse shape, the materials used in its construction, orientation, and management systems can have a large impact on the greenhouse's performance. Therefore, optimizing greenhouses is truly important to ensure their contribution to the sustainability of food production.

The evolution of computational power allows numerical modeling to be used to study these structures while also keeping cost and resource requirements low, when compared with developing physical prototypes. Empirical evidence is, nevertheless, a requirement in numerical modeling as a means to assess the reliability of the simulation results. Recent studies show that various building characteristics can have a significant impact on the energy required to keep the conditions inside a greenhouse suitable for crop growth. An optimization procedure integrating a dynamic thermal model that was developed for optimal design of solar greenhouses in different climate conditions revealed that an optimal solar greenhouse can work passively 85% of the time over a year without additional energy sources [4].

The shape of the roof of the greenhouse can reduce the heating demand by up to 4.2% [5]. East–west orientation can lower energy requirements of a greenhouse operating year-round [6]. The usage of a greenhouse cover with higher insulating properties was able, through computer simulations in EnergyPlus, to decrease energy consumption by more than 30% [7]. Ansys Fluent simulations also demonstrated the benefits of carefully selecting the greenhouse cover [8]. The various design characteristics of a greenhouse that result in higher efficiency in the situation under study are not always applicable to every other greenhouse that is being optimized, such as the orientation of the structure, the optimal value of which varies with the latitude of the site where the greenhouse will be implemented [9,10]

The microclimates that develop around and between the crop canopies require very detailed modeling and, consequently, high computational power. However, there has been success in replicating these thermal interactions in numerical models [11], with simulations of microclimates being useful to optimize plant pot position and frequency of movement, resulting in a reduction by 90% in microclimate formation and a 95% reduction in frequency of pot movement [12]. To obtain numerical model results that better match the real-world conditions experienced inside a greenhouse during crop development, it has been shown that modeling not only the structural components of the greenhouse but also the crops themselves is a valuable aspect of the simulation process [13,14].

While existing computer software has shown to be capable of assessing the thermal behavior of a greenhouse, research has been done to develop tools that are capable of outputting better results, with an application based on TRNSYS showing discrepancies of 50 kWh/m^3 comparatively with traditional EnergyPlus simulations [15]. Mathematical models that were developed for specific greenhouse assessment and subsequently solved with the computer application MATLAB/Simulink have also demonstrated ability to correctly estimate the thermal performance of a greenhouse [16].

In this paper, a computational model of an existing small-scale greenhouse was built using the computer software EnergyPlus and DesignBuilder. The model was then simulated under the same conditions that were applied to the physical greenhouse to evaluate the reliability of the simulation results. Subsequently, a group of case studies was defined to evaluate the impact of different design choices in the thermal behavior of the greenhouse.

2. Materials and Methods

2.1. Experimental Setup

The greenhouse under study is located in the city of Covilhã at the University of Beira Interior Engineering Faculty campus in the center-east region of Portugal, having the following GPS coordinates: 40°16′43″ N 7°30′48″ W.

The greenhouse has a rectangular base with a traditional gable roof, with exterior dimensions of 2.0 m × 1.4 m × 2.1 m (length × width × height). Figure 1 shows in more detail the supporting structure of the greenhouse and its dimensions.

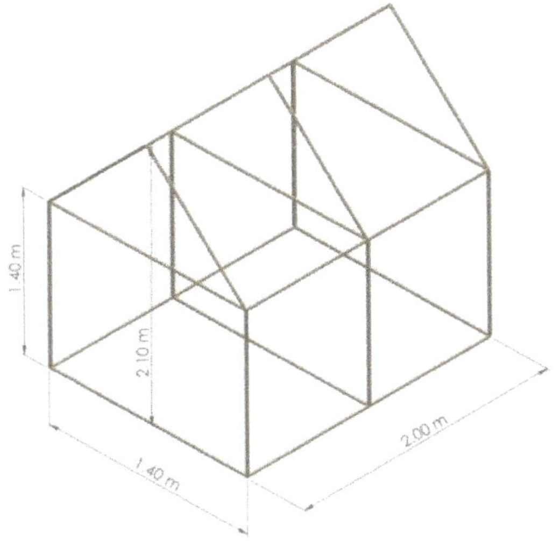

Figure 1. External dimensions of the greenhouse under study.

The supporting structure is made from fir wood, with cross-sectional dimensions of 30 mm × 30 mm. The cover, which involves all sides of the greenhouse, is made of a commercial polyethylene 215 μm in thickness commonly used for this kind of greenhouse. Figure 2 shows the greenhouse with the covering material assembled.

Figure 2. Picture of the greenhouse with the polyethylene cover applied.

The greenhouse has two circular openings of 127 mm in diameter located in the bottom-left corner of the south-facing wall and the top-right corner of the north-facing wall to allow natural or forced ventilation of the interior of the greenhouse. Figure 3 shows one of these openings with a mechanical ventilator duct attached.

Figure 3. Picture of one of the circular openings of the greenhouse with a ventilation duct attached.

The greenhouse location is surrounded by various buildings and a retaining wall that are tall enough to influence its thermal performance by blocking sunlight during certain periods of the day.

Temperature and relative humidity data acquisitions were performed by means of two devices, the PCE-T 1200 temperature data logger, using T-type thermocouples (Omega Engineering Inc., Norwalk, CT, USA) and the Lascar Electronics EL-USB-1-LCD portable data logger. The accuracies of the temperature and relative humidity measurements were ±0.55 °C and 2.25%, respectively. In order to ensure stable thermal conditions inside the greenhouse, it was necessary for some tests to use a heating device. The equipment used was a ceramic fan heater with a digital thermostat and maximum power of 1800 W.

The experimental type A tests included natural and forced ventilation inside the greenhouse. Some tests (type B) included a heating set-point temperature to ensure a minimum value of the internal air temperature by using an auxiliary heater located in the greenhouse. Type C tests included a low airflow rate inside the greenhouse with outside fresh air but keeping the auxiliary air heater turned off. Table 1 lists the parameters of each of the experimental tests. The experimental tests were each conducted over 24 h, and the air temperature and relative humidity were recorded every 10 min. Additional details of the experimental apparatus and experimental tests can be found in [17,18].

Table 1. Experimental test parameters necessary for model validation [17,18].

Tests	Ventilation Type	Airflow Rate (ac/h)	Auxiliary Heating	Heating Set-Point (°C)
A1	Natural	-	OFF	-
A2	Forced	12.9	OFF	-
A3	Forced	14.7	OFF	-
A4	Natural	-	OFF	-
B5	Natural	-	ON	12
B6	Natural	-	ON	20
B7	Forced	12.9	ON	18
C1	Forced	1.0	OFF	-
C2	Forced	3.0	OFF	-
C3	Forced	3.0	OFF	-

2.2. Computational Model

The computer application EnergyPlus version 8.1 with graphical user interface DesignBuilder version 3.4.041 was used to simulate the dynamic thermal behavior of the greenhouse model. This well-known computational tool was developed and applied mostly for more traditional structures, such as residential, office, and commercial buildings [19–23]. However, it has been used successfully in simulating considerably different structures from the aforementioned ones, such as data centers, outdoor telecommunications cabinets, or greenhouses [19,24,25].

Figures 4 and 5 show the model of the greenhouse under study and its surrounding structures in order to include the shadow effects on the greenhouse during certain periods of the day.

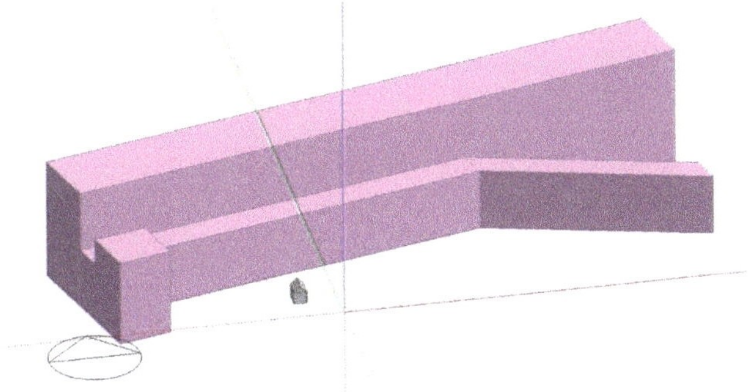

Figure 4. Visualization of the greenhouse model (in gray) and surrounding structures (in purple).

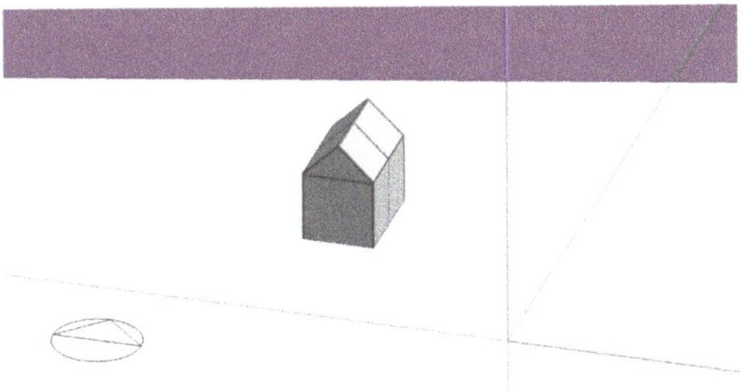

Figure 5. Close-up of the model of the greenhouse.

With the model's geometry defined, the following step was to set the building material's characteristics in the model. As such, the extensive library of material templates included with DesignBuilder was used for most of the greenhouse's materials. For the supporting wood structure, the template "Woods-fir, pine" was selected and modified to have a thickness of 0.03 m. This template was applied to the "External walls", "Pitched roof (occupied)", and "Internal partitions" sections of the model.

The floor of the greenhouse is made from the natural terrain found in the greenhouse's location, and so the template "Earth, common (0.5 m)" was chosen. Air infiltration was considered in this model. Although no experimental data were available—as it is a small greenhouse built with special care for research purposes and located in a place relatively sheltered from the wind—an infiltration value of 0.2 ac/h (air chamber volume changes per hour) was considered. The properties of the surrounding buildings are also relevant, and so the "Project component block material" template was set for these structures. Table 2 summarizes the characteristics of the polyethylene cover, which were inputted manually, with the thermal conductivity being sourced from the ISO 10456 standard and the other necessary properties from Kempkes et al. [18].

Table 2. Polyethylene cover properties used for the model.

Property	Value
Thickness (mm)	0.215
Thermal conductivity (W/m·K)	0.330
Solar transmittance	0.800
Outdoor solar reflectance	0.170
Inside solar reflectance	0.170
Visible transmittance	0.850
Outdoor visible reflectance	0.120
Inside visible reflectance	0.120
Infra-red transmittance	0.000
Outdoor emissivity	0.700
Inside emissivity	0.700

A climatic data file containing the weather data of Covilhã was used [26] for simulation purposes. This data file includes 2002 as the reference year. Furthermore, a "10 time steps per hour" setting was adopted, and the outdoor ambient air temperature was used as the criterion to determine which simulation day more closely matched the day of each experimental test. Additional details of the greenhouse modeling can be found in [27].

3. Results

This section includes the comparison of the results of the computational tool with the results obtained from a real greenhouse, followed by a detailed set of case studies.

3.1. Comparison with Experimental Data

Figure 6 shows the results of the simulation interval of 28 May at 17:00 to 29 May at 16:00 for the test A1, specifically the outdoor and indoor air temperatures of the greenhouse for both the experimental and the simulated results.

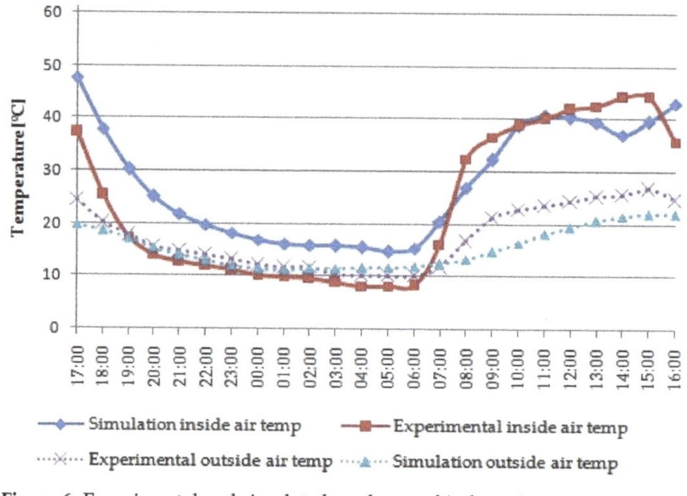

Figure 6. Experimental and simulated outdoor and indoor air temperatures in test A1.

There was considered to be natural ventilation with an airflow of 0.2 ac/h without auxiliary heating for the dynamic simulation of the greenhouse in test A1. The trend of the indoor air temperature is similar in both numerical and experimental tests. However, the numerical predictions are above the experimental values during the night period, showing an air temperature difference between 7 °C and 10 °C. On the other hand, during the day the temperature predicted for the indoor air of the greenhouse is almost always above the value measured experimentally, although the temperature of the outdoor air measured at the site is higher than that considered for modeling purposes. It reached a maximum difference of 6.5 °C early in the day.

The comparison results for the period between 29 May at 17:00 and 30 May at 16:00, test B7, are given in Figure 7. This test includes an airflow rate of 12.9 ac/h and an auxiliary air heater with a set-point of 18 °C. The comparison shows a similar trend of the numerical and experimental indoor air temperature values. During the day, when the numerical values of the outdoor air temperature are lower than the experimental ones, a similar behavior is observed, but a temperature difference of about 10 °C is reached between numerical and experimental results of the indoor air temperature. This trend arises from the role of the airflow rate, which affects the indoor air temperature in both situations.

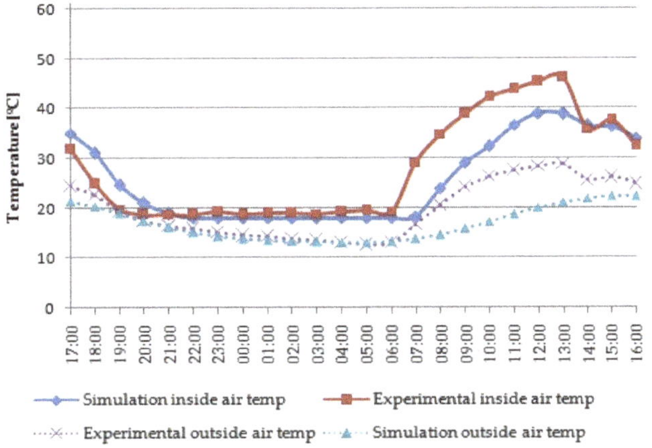

Figure 7. Test B7 results for outdoor and indoor air temperatures.

Figure 8 shows comparison results for the period between 25 February at 17:00 and 26 February at 16:00, test C3. In this test, there was considered to be an airflow rate of 3 ac/h without auxiliary air heater. Again, a similar trend was found. The lower value of the indoor air temperature is reached at about the same time, around 06:00. During the day, the outdoor air temperature difference between the experimental and numerical models reaches a maximum value of about 16 °C, resulting in lower values for the predictions of the indoor air temperature by about 10 °C.

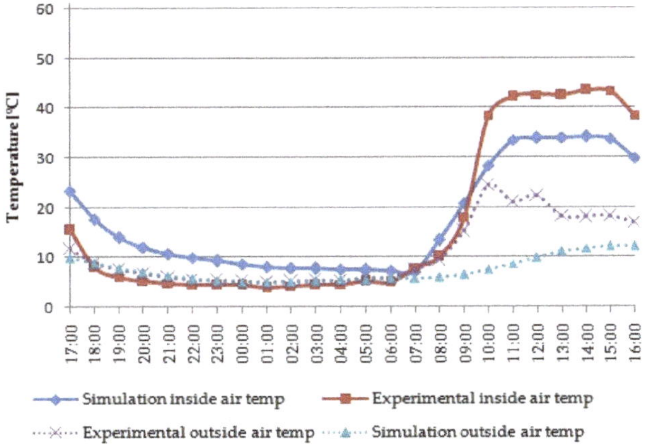

Figure 8. Experimental air temperature and numerical predictions for test C3.

Overall, the comparison results show that the computational tool can provide a reasonable prediction of the indoor air temperature of the greenhouse. However, it must pointed out that the climatic data files used in this simulation software could significantly influence the quality of the results, as observed by other authors [28–30].

3.2. Case Studies

Taking advantage of the potential of the computational tool employed, a selection of case studies was defined in order to understand and quantify the consequences for the greenhouse thermal behavior resulting from the change of some individual characteristics of its envelope. In all cases studies, the model equivalent to the experimental greenhouse is always taken as the reference case.

3.2.1. Geometry and Orientation

The first group of case studies analyzes the thermal behavior of a greenhouse with different overall dimensions while maintaining the same floor area and total height, when applicable, as well as the impact of the greenhouse orientation.

Figure 9 shows the first case study, in which a greenhouse with a cross-sectional profile of a quarter circle was adopted.

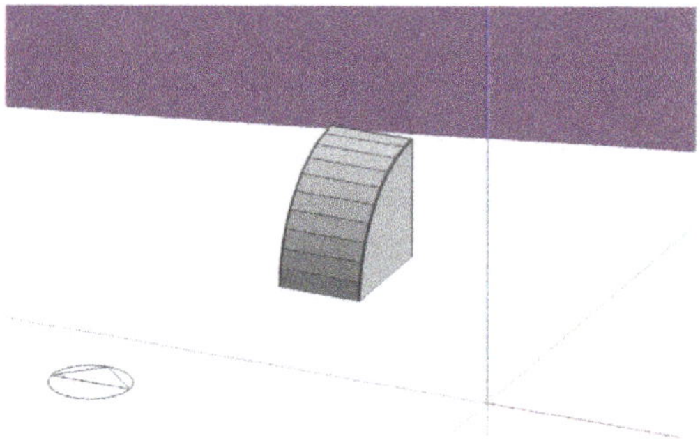

Figure 9. Model of the greenhouse—case study 1.

Case study 2 implemented a cross-sectional profile of a semicircle, resulting in a greenhouse with a considerably smaller internal volume, which can be seen in Figure 10.

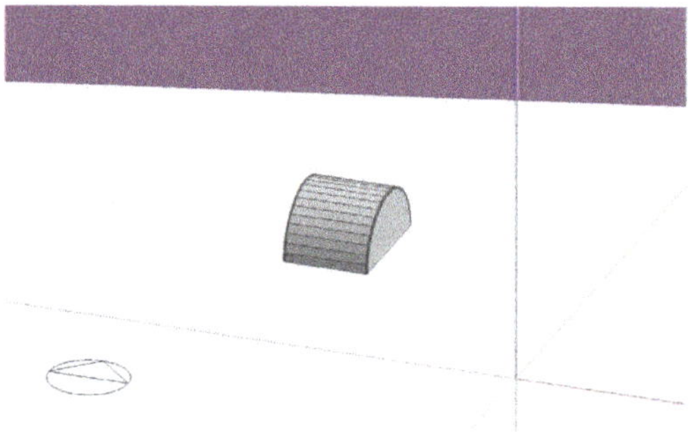

Figure 10. Model of the greenhouse—case study 2.

The third case study used a similar geometry to case study 2, where the geometry of case study 2 was raised so that it had the same height at the highest point as the reference greenhouse. Figure 11 shows the third case study configuration.

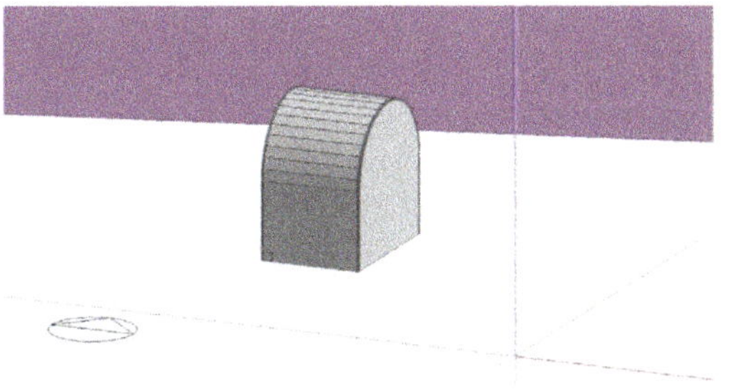

Figure 11. Model of the greenhouse—case study 3.

In order to assess the effect of greenhouse orientation, in the last case study of this group the reference greenhouse, keeping the same dimensions, is simply rotated 90° counterclockwise, so that the initially south-facing wall is now oriented to the east, as shown in Figure 12.

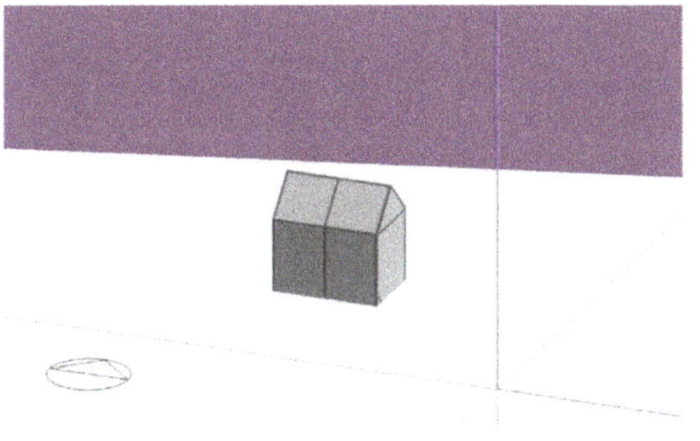

Figure 12. Model of the greenhouse—case study 4.

The annual results of this group of case studies can be found in Figures 13 and 14, which show, respectively, the indoor air temperature and relative humidity for the reference greenhouse as well as for the four case studies. The outdoor air temperature is also shown in both figures.

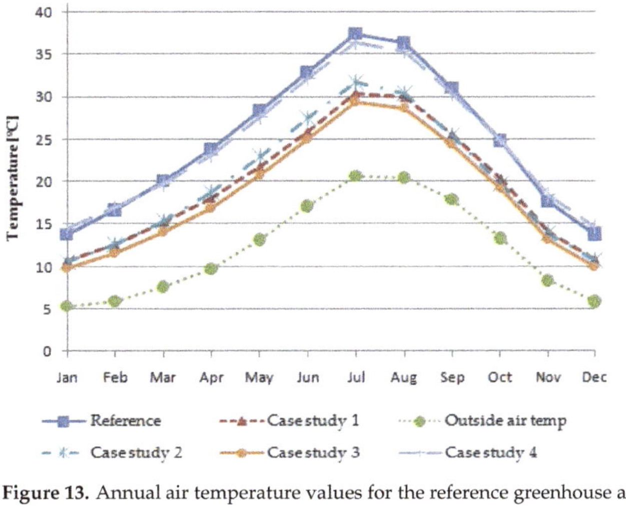

Figure 13. Annual air temperature values for the reference greenhouse and case studies 1–4.

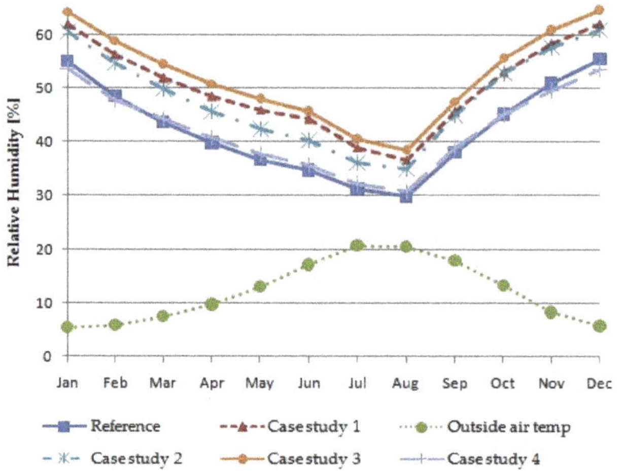

Figure 14. Annual relative humidity values for the reference greenhouse and case studies 1–4.

The results show that the different configurations tested have a worse thermal performance than the reference case. In the worst case, case study 3, the reduction in indoor air temperature goes from around 5 °C in January to 8 °C in July. Changing the orientation of the greenhouse—case study 4—makes it possible to slightly increase the indoor air temperature by around 0.8 °C in January and reduce it in July by around 1 °C.

The annual air relative humidity results show a similar but symmetrical behavior to that of the annual air temperature. The highest indoor air relative humidity of 64% is obtained in case study 3, which is 11% above the reference case, in January. The minimum air humidity value is obtained in August in the reference case with a value of 31%, which is 8% below case study 3. It can be concluded that the reference case with the orientation provided by case study 4 is the better choice in terms of greenhouse indoor air temperature behavior.

3.2.2. Airflow Rate

The second group of case studies evaluates the effect of increasing the rate of the renovation of the air inside the greenhouse by means of mechanical ventilation. The airflow rate values simulated were 1 ac/h, 2 ac/h, 3 ac/h, 4 ac/h, and 5 ac/h for case studies 5 through 9, respectively. Figures 15 and 16 show the air temperature values and relative humidity values, respectively, for the second group of case studies for the simulation day of 15 January. The choice of this day is related to the fact that it is the most demanding in terms of energy consumption to provide the interior heating of the greenhouse.

As shown in Figure 16, the relative humidity values show that the increase in airflow rate resulted in a steady increase in relative humidity during the day as well as the night, with an exception in the later hours of the night when the reference greenhouse's relative humidity was about 10% higher than any of the case studies as a result of a much lower airflow rate. It should be mentioned that in a real greenhouse, the relative humidity of the indoor air during the day would never be so low, due to the products breathing. The results show that the variation of the airflow rate allows adjustment of the temperature of the indoor air in the greenhouse, which, even on a winter day, can reach temperatures close to 30 °C at certain times of the day. A more stable relative humidity is also achievable by increasing the airflow rate.

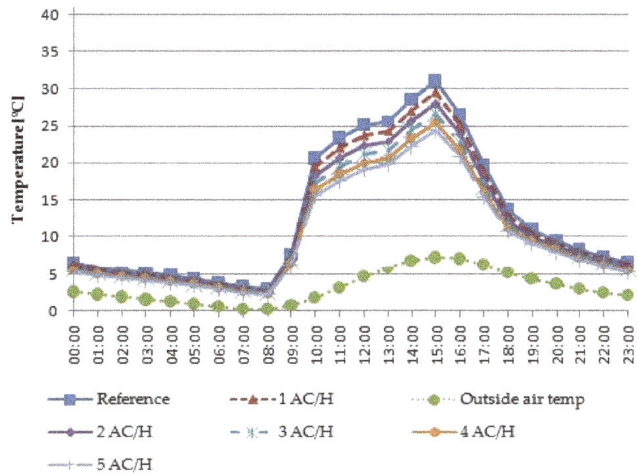

Figure 15. Air temperature values for case studies 5–9 on 15 January.

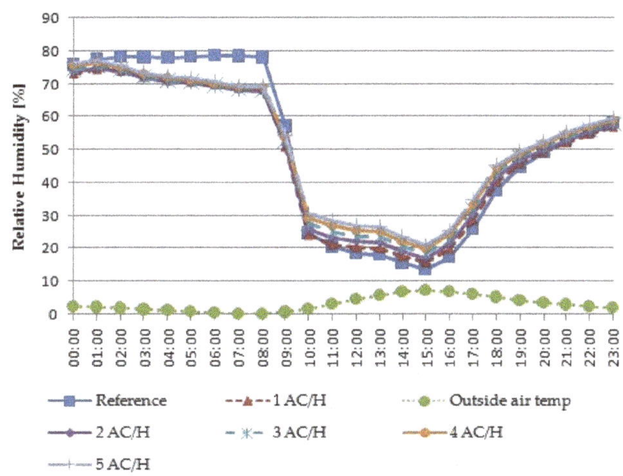

Figure 16. Relative humidity values for case studies 5–9 on 15 January.

As expected, the results show that increasing the outdoor air input results in lower temperatures during the day, reaching an air temperature difference of 6.5 °C between the 0.2 ac/h case (reference case) and the 5 ac/h case (see Figure 15). The minimum air temperature is obtained at 08:00 and varies between 2.2 °C and 2.9 °C, for the highest and lowest airflow rate, respectively. During the night, there is a reduced impact of the airflow rate on the variation of the temperature inside the greenhouse, due to the proximity between the indoor and outdoor temperatures.

3.2.3. Greenhouse Cover

Two case studies were carried out to study the effect of the characteristics of the greenhouse cover on its thermal performance. In case study 10, the greenhouse envelope is made up of two polyethylene film separated by a 13 mm layer of air. In case study 11, three polyethylene films are used instead of two but with the same thickness of air separation between them.

The air temperatures inside the greenhouse for the different types of cover over the year are given in Figure 17. The three-layer cover solution allows obtention of the highest temperature of the greenhouse's indoor air throughout the year, surpassing the reference case, which has a simple cover, by 3.5 °C in January and 7.3 °C in July. However, the two-layer setup appears to be more cost-effective and does not excessively penalize the air temperature reached, with a temperature reduction of just 1 °C in January.

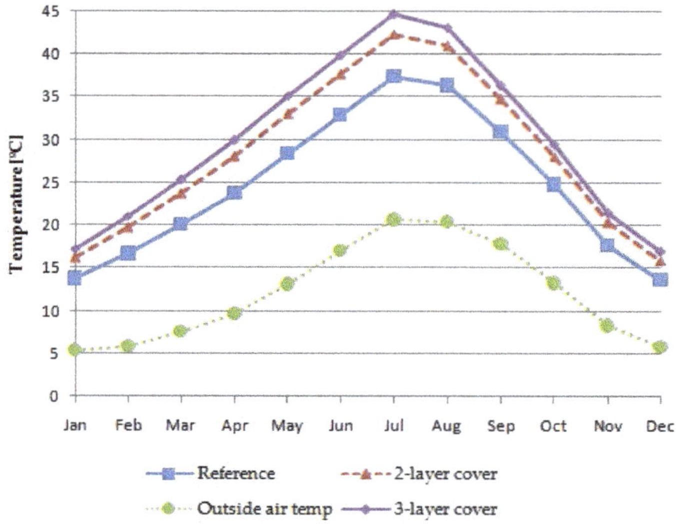

Figure 17. Annual air temperature values for the reference greenhouse and case studies 10 and 11.

The relative air humidity results of case studies 10 and 11 are shown in Figure 18. The two- and three-layer configurations are capable of significantly reducing the relative humidity inside the greenhouse throughout the year, due to the higher air temperature reached. These two case studies allow us to conclude that the quality of the greenhouse cover also plays a relevant role in the thermal conditions reached inside the greenhouse.

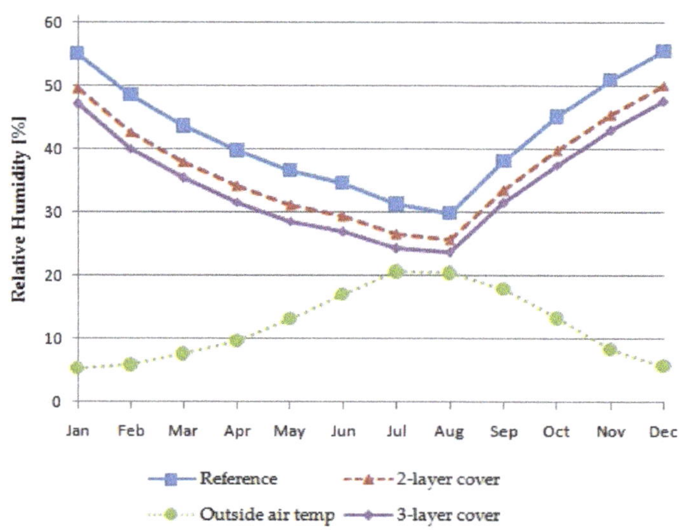

Figure 18. Annual relative humidity values for the reference greenhouse and case studies 10 and 11.

3.2.4. Outer-Facing Side Cover Emissivity

The fourth group of case studies focuses on the emissivity value of the outer-facing side of the greenhouse cover. Case studies 12 through 15 change this parameter to 0.1, 0.3, 0.5, and 0.9, respectively. Figures 19 and 20 contain the temperature and relative humidity results for these case studies.

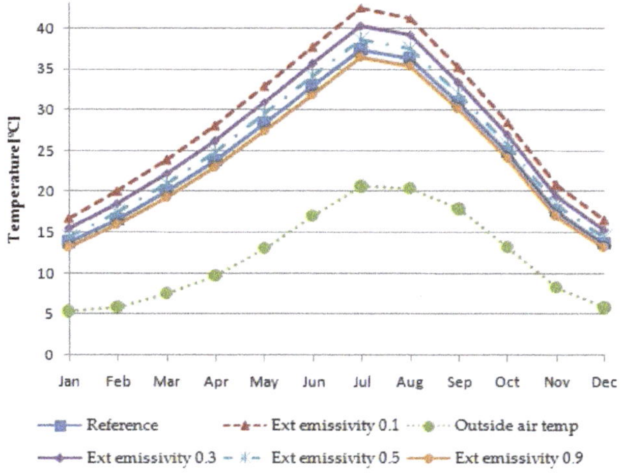

Figure 19. Annual air temperature values for the reference greenhouse and case studies 12–15.

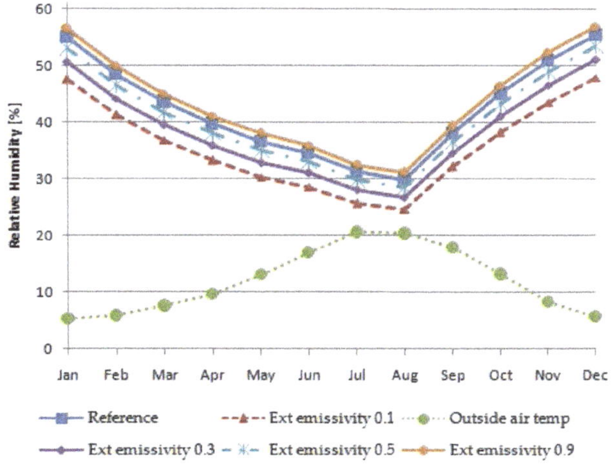

Figure 20. Annual relative humidity values for the reference greenhouse and case studies 12–15.

As expected, the difference of relative humidity inside the greenhouse between the reference case and the lowest emissivity case decreased is 7.4% in January and 5.0% in July, due to the correlation between air temperature and relative air humidity. See Figure 20 for the variation of relative humidity during the year.

This group of case studies shows that by decreasing the emissivity value on the outside of the polyethylene cover, there is an increase in the indoor air temperature of the greenhouse, reaching 2.9 °C in January and 5 °C in July for an emissivity value of 0.1 in relation to the reference case. This indoor air temperature increase is due to the reduction in infrared radiation escaping from the interior through the cover.

3.2.5. North-Facing Wall Material and Thickness

In order to evaluate the role of the thermal mass in the thermal performance of the greenhouse, solid granite rock walls with 40 cm, 60 cm, and 100 cm thicknesses were considered for the north-facing wall. The "grounded" boundary condition was also applied, meaning the outer edge of the granite is in contact with ground terrain/soil instead of outdoor ambient air.

The results of indoor air temperatures of the greenhouse for these case studies are shown in Figures 21 and 22, respectively, for the middle days of January and July.

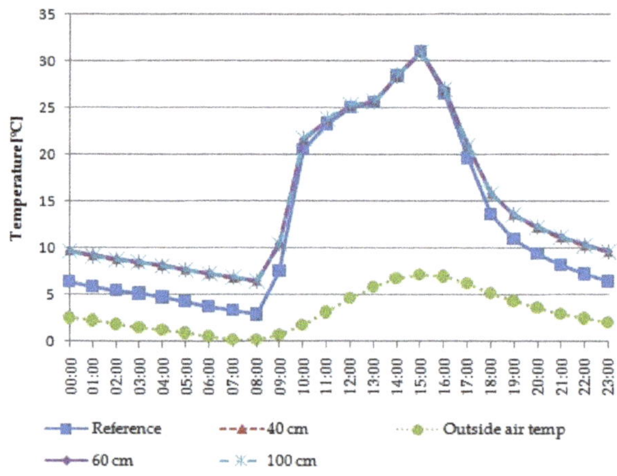

Figure 21. Indoor air temperature on 15 January for different wall thicknesses.

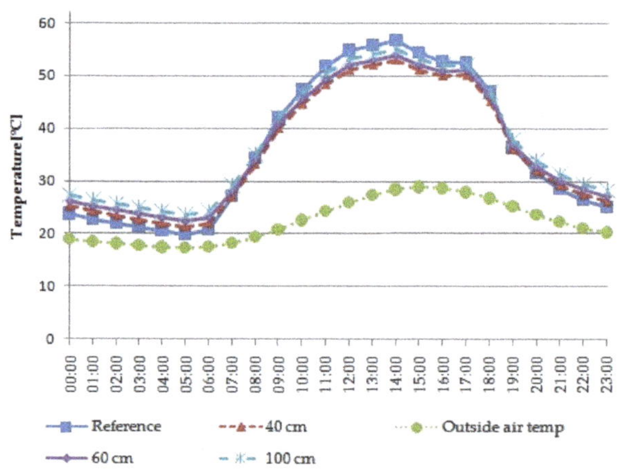

Figure 22. Indoor air temperature on 15 July for different wall thicknesses.

The July values show that the wall thickness affects the obtained air temperatures, with the minimum value, at 05:00, increasing from 21.4 °C to 23.7 °C when the wall thickness increases from 40 to 100 cm. At 14:00, the maximum indoor air temperature of the greenhouse shows a difference of 2 °C between the maximum and minimum wall thicknesses but, in any case, below the temperature of 56.7 °C of the reference case at the same time. In January, the thermal mass of the granite wall allows the minimum indoor air temperature of the greenhouse, reached at 08:00, to be 3.6 °C above the reference greenhouse and 6.3 °C above the outdoor temperature regardless of the thickness considered for the wall. The maximum indoor air temperature reached at 15:00 is similar to the reference greenhouse's value and is also unaffected by the thickness of the granite wall, remaining at 23.7 °C above the outdoor ambient temperature at that time.

Overall, the air temperature values show the ability of the high thermal inertia of the large granite wall to act as a heat reservoir that releases heat during the colder periods and absorbs heating during the hotter ones. However, it also shows that the wall thickness should be balanced with a careful airflow rate to limit the maximum air temperatures reached.

3.2.6. Geographical Location

The final group of case studies consists in simulating the reference greenhouse in various locations of the globe so as to assess its performance and limitations in different climates. Case studies 19 to 21 have the reference greenhouse set in Lisbon, Moscow, Ostersund, and Riyadh, respectively. Figures 23 and 24 show the results of these case studies for the day of 15 January, and Figures 25 and 26 contain the results for the day of 15 July.

As shown in Figure 24, the relative humidity values for 15 January demonstrate that Moscow and Riyadh have a more constant relative humidity throughout the day, although reaching relatively high and low values, respectively. Ostersund and Lisbon have a wider range of values, with Lisbon having bigger oscillations but not surpassing the highest and lowest values of all case studies.

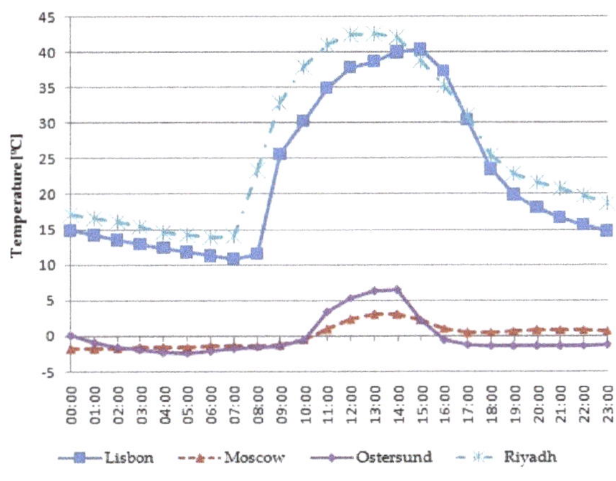

Figure 23. Air temperature values for case studies 19–21 on 15 January.

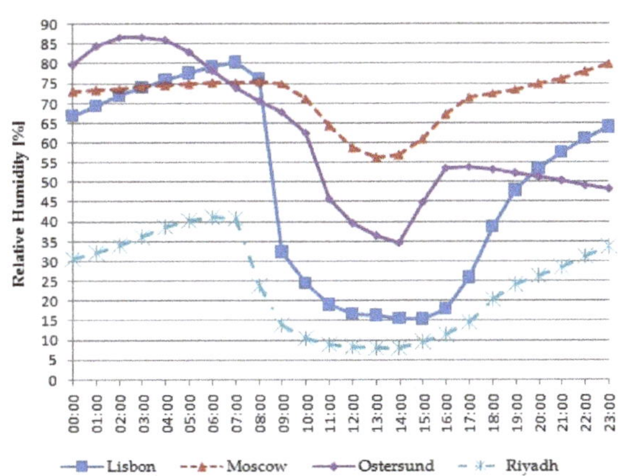

Figure 24. Relative air humidity values for case studies 19–21 on 15 January.

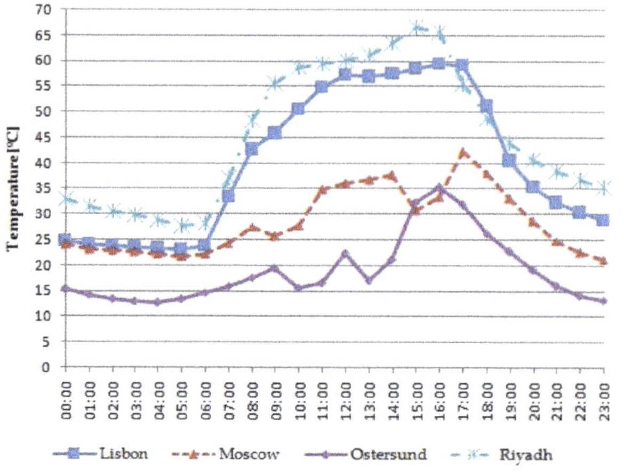

Figure 25. Air temperature values for case studies 19–21 on 15 July.

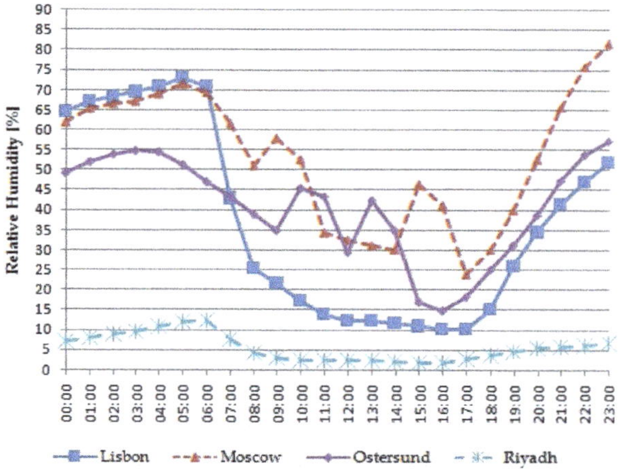

Figure 26. Relative air humidity values for case studies 19–21 on 15 July.

The temperature results in January show that the greenhouse mostly requires heat extraction during the day when located in Lisbon and Riyadh in order to achieve an adequate temperature for operation. When located in Moscow and Ostersund, constant auxiliary heating is necessary to achieve an appropriate temperature in the greenhouse. Thus, this greenhouse is better suited for warmer climates if winter operation is required.

The air temperature results for 15 July shown in Figure 25 allow us to conclude that the greenhouse reaches very high temperatures in Lisbon and, especially, Riyadh, meaning that constant heat extraction will be necessary to achieve adequate temperature levels for any kind of crop development. Moscow's and Ostersund's temperature values show that the greenhouse is very suitable for these locations, as it able to maintain a much more constant and adequate temperature throughout the day.

As shown in Figure 26, the relative humidity values for case studies 19 to 21 show the greenhouse, when located in Riyadh, having a very constant and low relative humidity level, followed by Ostersund, where a larger range but average relative humidity values are observed, and, lastly, with Lisbon and Moscow showing the largest oscillation as the day progresses. This set of case studies shows that the reference greenhouse's capabilities depend not only on the location but also on the season. The reference greenhouse's low insulation mean that additional energy sources will be necessary to provide more adequate ambient conditions, particularly if the greenhouse is to be operated during the entire year.

3.2.7. Greenhouse Energy Consumption

In this last case study, an assessment of the energy consumption of the greenhouse throughout the year was carried out after the introduction of the improvements recommended in the previous subsections. Specifically, the greenhouse was reoriented by 90°; a double cover with 13 mm air layer was used; the emissivity of the inside cover was set to 0.3; a granite wall of 40 cm thickness on the north wall was included; and a variable airflow rate, according to the schedule indicated in Figure 27, with a maximum value of 5 ac/h was used. For the greenhouse indoor thermal conditions, a set-point of 21 °C was used, suitable for the production of the most consumed vegetables [31].

Month	Monday	Tuesday	Wednesday	Thursday	Friday	Saturday	Sunday
Jan	Off	Off	Off	Off	Off	Off	Off
Feb	Off	Off	Off	Off	Off	Off	Off
Mar	10:00 - 17:00 A 40PER	10:00 - 17:00 A 40PER	10:00 - 17:00 A 40PER	10:00 - 17:00 A 40PER	10:00 - 17:00 A 40PER	10:00 - 17:00 A 40PER	10:00 - 17:00 A 40PER
Apr	10:00 - 17:00 A 40PER	10:00 - 17:00 A 40PER	10:00 - 17:00 A 40PER	10:00 - 17:00 A 40PER	10:00 - 17:00 A 40PER	10:00 - 17:00 A 40PER	10:00 - 17:00 A 40PER
May	9:00 - 18:00 A 80PER	9:00 - 18:00 A 80PER	9:00 - 18:00 A 80PER	9:00 - 18:00 A 80PER	9:00 - 18:00 A 80PER	9:00 - 18:00 A 80PER	9:00 - 18:00 A 80PER
Jun	9:00 - 18:00 A 80PER	9:00 - 18:00 A 80PER	9:00 - 18:00 A 80PER	9:00 - 18:00 A 80PER	9:00 - 18:00 A 80PER	9:00 - 18:00 A 80PER	9:00 - 18:00 A 80PER
Jul	8:00 - 19:00 A 100PER	8:00 - 19:00 A 100PER	8:00 - 19:00 A 100PER	8:00 - 19:00 A 100PER	8:00 - 19:00 A 100PER	8:00 - 19:00 A 100PER	8:00 - 19:00 A 100PER
Aug	8:00 - 19:00 A 100PER	8:00 - 19:00 A 100PER	8:00 - 19:00 A 100PER	8:00 - 19:00 A 100PER	8:00 - 19:00 A 100PER	8:00 - 19:00 A 100PER	8:00 - 19:00 A 100PER
Sep	9:00 - 18:00 A 80PER	9:00 - 18:00 A 80PER	9:00 - 18:00 A 80PER	9:00 - 18:00 A 80PER	9:00 - 18:00 A 80PER	9:00 - 18:00 A 80PER	9:00 - 18:00 A 80PER
Oct	Off	Off	Off	Off	Off	Off	Off
Nov	Off	Off	Off	Off	Off	Off	Off
Dec	Off	Off	Off	Off	Off	Off	Off

Figure 27. Annual airflow rate schedule.

Additionally, for comparative purposes, a variable airflow rate schedule and temperature set-point of 21 °C were also used in the reference greenhouse. Figure 28 illustrates the results for two locations with very different characteristics: Lisbon and Ostersund. While a greenhouse located in Ostersund requires an energy consumption of 691.5 kWh in January, the same greenhouse located in Lisbon requires only 162.2 kWh to ensure the same indoor thermal conditions. The result is about three times less. Regarding the reference greenhouse, the changes introduced allowed the energy consumption for the month of January to be reduced by 57% in Lisbon and 43% in Ostersund. Even in July, it was possible to reduce the average energy consumption in Ostersund from 83.4 kWh to 28.9 kWh.

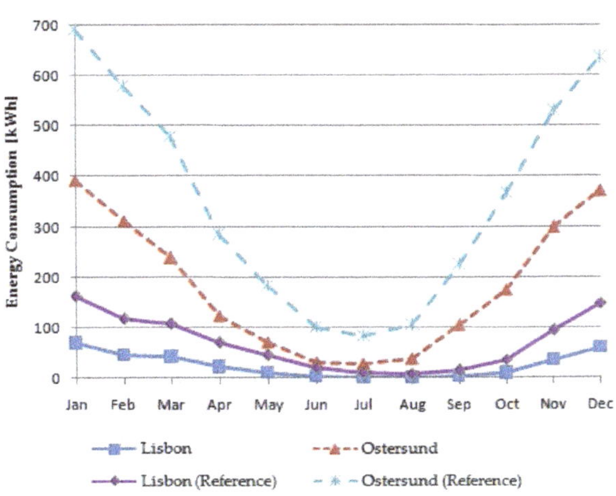

Figure 28. Annual energy consumption values for the greenhouses.

4. Conclusions

The results obtained from the simulation model developed in this paper have shown the effectiveness of EnergyPlus and DesignBuilder in assessing the thermal behavior of a greenhouse, as the air temperature inside the greenhouse followed a similar overall pattern when subject to approximately the same outside weather conditions. The case studies showed that the reference case configuration presents better thermal behavior for the same floor area. In the worst case, case study 3, the reduction in indoor air temperature goes from around 5 °C in January to 8 °C in July. Additionally, the east–west orientation allows a slight increase of the indoor air temperature by around 0.8 °C in January and reduction in July by around 1 °C. The number of greenhouse air changes per hour affects the indoor thermal conditions. The results for 15 January show that increasing the outdoor air input results in lower temperatures during the day, reaching an air temperature difference of 6.5 °C between the 0.2 ac/h case (reference case) and the 5 ac/h case. During the night, there is a reduced impact of the airflow rate on the variation of the temperature inside the greenhouse, due to the proximity between the indoor and outdoor temperatures. Better insulation, through lower emissivity covering material and usage of multiple layers of covering material, is capable of greatly increasing the greenhouse's solar energy absorption, resulting in much lower heating demands in colder regions and during the winter season. A three-layer cover solution of polyethylene film with an air gap of 13 mm allows obtaining the highest temperature of the greenhouse indoor air throughout the year, surpassing the reference case with a simple cover by 3.5 °C in January and 7.3 °C in July. The indoor air temperature of the greenhouse increases as well, reaching 2.9 °C in January and 5 °C in July, when the emissivity of the cover is reduced to 0.1 in relation to the reference case. The usage of a granite north wall can result in a considerably more constant temperature inside the greenhouse, due to its high thermal mass, although the sizing of this wall should be balanced with a careful airflow rate to limit the maximum air temperatures reached. In January, the minimum indoor air temperature of the greenhouse, reached at 08:00, is 3.6 °C above the reference greenhouse and 6.3 °C above the outdoor temperature regardless of the thickness considered for the wall. However, the July values show that the wall thickness affects the minimum and maximum values of the indoor air temperatures by about 2 °C when the wall thickness increases from 40 to 100 cm. If the aforementioned changes are introduced together in the greenhouse, the results for two locations with very different characteristics, Lisbon and Ostersund, show that a greenhouse located in Lisbon requires about three times less energy consumption to achieve the same indoor thermal conditions in January. Furthermore, the changes introduced allowed the energy consumption for the month of January to be reduced by 57% in Lisbon and by 43% in Ostersund with respect to the reference greenhouse.

Author Contributions: Conceptualization, P.D.S. and L.C.P.; methodology, P.D.S.; validation, P.D.S., L.C.P. and B.L.; formal analysis, P.D.S. and B.L.; investigation, P.D.S. and B.L.; resources, P.D.S., L.C.P. and P.D.G.; writing—original draft preparation, B.L.; writing—review and editing, P.D.S., L.C.P. and P.D.G.; visualization, P.D.S. and B.L.; supervision, P.D.S.; project administration, P.D.S., L.C.P. and P.D.G. All authors have read and agreed to the published version of the manuscript.

Funding: This research received no external funding.

Institutional Review Board Statement: Not applicable.

Informed Consent Statement: Not applicable.

Data Availability Statement: The data used to support the findings of this study are available from the corresponding author upon request.

Acknowledgments: P.D.G. and P.D.S. acknowledge Fundação para a Ciência e a Tecnologia (FCT—MCTES) for its financial support via the project UIDB/00151/2020 (C-MAST).

Conflicts of Interest: The authors declare no conflict of interest.

References

1. Abderrahman, M.; Walid, R.; Ammar, M.; Abdelaziz, B.; Abdelkader, O. Optimum design and orientation of a greenhouse for seasonal win ter drying in Morocco under constant volume constraint. *Sol. Energy* **2021**, *230*, 321–332.
2. Khalid, A.J.; Ammar, A.F. A dynamic model and an experimental study for the internal air and soil temperatures in an innovative greenhouse. *Energy Convers. Manag.* **2015**, *91*, 76–82.
3. Sahdev, R.K.; Kumar, M.; Dhingra, A.K. A comprehensive review of greenhouse shapes and its applications. *Front. Energy* **2019**, *13*, 427–438. [CrossRef]
4. Esmaeli, H.; Roshandel, R. Optimal design for solar greenhouses based on climate conditions. *Renew. Energy* **2020**, *145*, 1255–1265. [CrossRef]
5. Gupta, M.J.; Chandra, P. Effect of greenhouse design parameters on conservation of energy for greenhouse environmental control. *Energy* **2002**, *27*, 777–794. [CrossRef]
6. Sethi, V.P. On the selection of shape and orientation of a greenhouse: Thermal modeling and experimental validation. *Sol. Energy* **2009**, *83*, 21–38. [CrossRef]
7. Fabrizio, E. Energy reduction measures in agricultural greenhouses heating: Envelope, systems and solar energy collection. *Energy Build.* **2012**, *53*, 57–63. [CrossRef]
8. Baxevanou, C.; Fidaros, D.; Bartzanas, T.; Kittas, C. Yearly numerical evaluation of greenhouse cover materials. *Comput. Electron. Agric.* **2018**, *149*, 54–70. [CrossRef]
9. Chen, C.; Li, Y.; Li, N.; Wei, S.; Yang, F.; Ling, H.; Yu, N.; Han, F. A computational model to determine the optimal orientation for solar greenhouses located at different latitudes in China. *Sol. Energy* **2018**, *165*, 19–26. [CrossRef]
10. Gupta, R.; Tiwari, G.N.; Kumar, A.; Gupta, Y. Calculation of total solar fraction for different orientation of greenhouse using 3D-shadow analysis in Auto-CAD. *Energy Build.* **2012**, *47*, 27–34. [CrossRef]
11. Mashonjowa, E.; Ronsse, F.; Milford, J.R.; Pieters, J.G. Modelling the thermal performance of a naturally ventilated greenhouse in Zimbabwe using a dynamic greenhouse climate model. *Sol. Energy* **2013**, *91*, 381–393. [CrossRef]
12. Ma, D.; Carpenter, N.; Maki, H.; Rehman, T.U.; Tuinstra, M.R.; Jin, J. Greenhouse environment modeling and simulation for microclimate control. *Comput. Electron. Agric.* **2019**, *162*, 134–142. [CrossRef]
13. Zhang, Y.; Henke, M.; Li, Y.; Yue, X.; Xu, D.; Liu, X.; Li, T. High resolution 3D simulation of light climate and thermal performance of a solar greenhouse model under tomato canopy structure. *Renew. Energy* **2020**, *160*, 730–745. [CrossRef]
14. Chahidi, L.O.; Fossa, M.; Priarone, A.; Mechaqrane, A. Energy saving strategies in a sustainable greenhouse cultivation in the mediterranean climate—A case study. *Appl. Energy* **2021**, *282*, 116156. [CrossRef]
15. Baglivo, C.; Mazzeo, D.; Panico, S.; Bonuso, S.; Matera, N.; Congredo, P.M.; Oliveti, G. Complete greenhouse dynamic simulation tool to assess the crop thermal well-being and energy needs. *Appl. Therm. Eng.* **2020**, *179*, 115698. [CrossRef]
16. Ali, R.B.; Bouadila, S.; Mami, A. Experimental validation of the dynamic thermal behavior of two types of agricultural greenhouses in the Meditterranean context. *Renew. Energy* **2020**, *147*, 118–129.
17. Godinho, N. Utilização de Permutadores de Calor ar-Solo na Climatização de Estufas Agrícolas. Master's Thesis, University of Beira Interior, Covilhã, Portugal, 2019.
18. Vitale, Z. Earth-to-Air Heat Exchanger Application in Passive Greenhouse Heating or Cooling for Agricultural Use: A practical Case. Master's Thesis, University of Beira Interior, Covilhã, Portugal, 2020.
19. Silva, P.D.; Pires, L.; Patrício, C.; Gaspar, P.D. Characterization of the thermal performance of an outdoor telecommunication cabinet. *Int. J. Energy Prod. Mgmt.* **2017**, *2*, 106–117. [CrossRef]
20. Kempkes, F.L.K.; Hemming, S. Calculation of NIR effect on greenhouse climate in various conditions. *Acta Hortic (Wagening)* **2012**, *927*, 543–550. [CrossRef]
21. Qiu, L.; Zhang, H.; Zhang, W.; Lai, D.; Li, R. Effect of existing residential renovation strategies on building cooling load: Cases in three Chinese cities. *Energy Build.* **2021**, *253*, 111548. [CrossRef]
22. Boyanoa, A.; Hernandez, P.; Wolfa, O. Energy demands and potential savings in European office buildings: Case studies based on EnergyPlus simulations. *Energy Build.* **2013**, *65*, 19–28. [CrossRef]
23. Al-Sakkaf, A.; Mohammed Abdelkader, E.; Mahmoud, S.; Bagchi, A. Studying Energy Performance and Thermal Comfort Conditions in Heritage Buildings: A Case Study of Murabba Palace. *Sustainability* **2021**, *13*, 12250. [CrossRef]
24. Pan, Y.; Yin, R.; Huang, Z. Energy modeling of two office buildings with data center for green building design. *Energy Build.* **2008**, *40*, 1145–1152. [CrossRef]
25. Liu, J.; Chen, Y.; Wang, B.; Wu, X.; Na, Y. Simulation and Experimental Study of Light and Thermal Environment of Photovoltaic Greenhouse in Tropical Area Based on Design Builder. *Appl. Sci.* **2021**, *11*, 10785. [CrossRef]
26. Aguiar, R.; Carvalho, M.J. Performance Analysis of Solar Thermal and Photovoltaic Systems. Solterm Software, National Laboratory of Energy and Geology Lisbon, Portugal, 2012. Available online: https://www.lneg.pt/en/service/technical-information/software/solterm5/ (accessed on 10 December 2021).
27. Lebre, B. Computational Modeling of the Thermal Behavior of a Greenhouse. Master's Thesis, University of Beira Interior, Covilhã, Portugal, 2021.
28. Chan, A. Developing future hourly weather files for studying the impact of climate change on building energy performance in Hong Kong. *Energy Build.* **2011**, *43*, 2860–2868. [CrossRef]

29. Bhandari, M.; Shrestha, S.; New, J. Evaluation of weather datasets for building energy simulation. *Energy Build.* **2012**, *49*, 109–118. [CrossRef]
30. Gonçalves, J.; Nunes, J.; Silva, P.D.; Gaspar, P.D.; Pires, L. Energy analysis and heat loads calculation approach: Application to agrifood industrial premises. In Proceedings of the 24th International Congress of Refrigeration, Yokohama, Japan, 16–22 August 2015.
31. Tüzel, Y.; Öztekin, G.B.; Gruda, N.; Sallaku, G.; Balliu, A.; Drăghici, E.M.; Brezeanu, P.M.; Muñoz, P.; Maršić, N.K.; Gvozdanovic-Varga, J.; et al. Good Agricultural Practices for greenhouse vegetable produtction in the South East European countries—Principles for sustainable intensification of smallholder farms. *FAO Plant Prod. Prot. Pap.* **2017**, *230*, 271–413.

applied sciences

MDPI

Article

Advances in Simulating Radiative Transfer in Complex Environments

Helge Simon *, Tim Sinsel and Michael Bruse

Department of Geography, Johannes Gutenberg University Mainz, 55099 Mainz, Germany; t.sinsel@geo.uni-mainz.de (T.S.); m.bruse@geo.uni-mainz.de (M.B.)
* Correspondence: h.simon@geo.uni-mainz.de; Tel.: +49-613-1392-7091

Abstract: Accurate simulation of radiative transfer is a very important aspect in climate modeling. For microclimate models in particular, it is not only important to simulate primary but also secondary radiative fluxes in great detail, i.e., emitted longwave and reflected shortwave radiation. As there are always limitations regarding computational effort and memory, these radiative fluxes are commonly implemented using simplified approaches. To overcome these simplifications and, thus, increase modeling accuracy, a new radiation scheme called indexed view sphere was introduced into the microclimate model ENVI-met. This new scheme actually accounts for radiative contributions of objects that are seen by each grid cell. In order to evaluate the advantages of the new scheme, it is compared against the formerly used averaged view factor scheme. The comparison in a complex realistic urban environment demonstrated that the indexed view sphere scheme improved the accuracy and plausibility of modeling radiative fluxes. It, however, yields an increased demand of memory to store the view facets for each cell. The higher accuracy in simulating secondary radiative fluxes should, however, overturn this shortcoming for most studies, as more detailed knowledge of local microclimatic conditions in general and eventually thermal comfort can be gained.

Keywords: ENVI-met; radiation scheme; microclimate; numerical modeling; thermal comfort; indexed view sphere; reflected radiation; longwave radiation; shortwave radiation; secondary radiative fluxes

Citation: Simon, H.; Sinsel, T.; Bruse, M. Advances in Simulating Radiative Transfer in Complex Environments. *Appl. Sci.* **2021**, *11*, 5449. https://doi.org/10.3390/app11125449

Academic Editors: María Isabel Lamas Galdo, Hassane Naji and Rodriguez J. Dios

Received: 10 May 2021
Accepted: 10 June 2021
Published: 11 June 2021

Publisher's Note: MDPI stays neutral with regard to jurisdictional claims in published maps and institutional affiliations.

Copyright: © 2021 by the authors. Licensee MDPI, Basel, Switzerland. This article is an open access article distributed under the terms and conditions of the Creative Commons Attribution (CC BY) license (https://creativecommons.org/licenses/by/4.0/).

1. Introduction

Accurate modeling of radiative fluxes plays an important role in microclimatology. This is especially the case in urban areas, where large differences in radiative fluxes can be found due to complex structures, heterogeneous materials, and a multitude of different surface types [1–5]. While primary radiative transfer, i.e., incoming shortwave and longwave radiation, can be simulated quite easily using ray tracing algorithms and local sky view factors, simulating secondary radiative fluxes, which are emitted or reflected by objects of the environment (walls, roofs, the ground surface, or vegetation), are much more complicated to be modeled. The complexity in modeling these radiative fluxes lies in the multiple interactions between the different elements within the view range of the grid analyzed. For instance, radiation reflected by a surface will contribute to the incoming radiation received by other surfaces. These surfaces will again also re-reflect parts of this radiation and distribute in vicinity. Handling such complex conditions is a challenge that is common to all algorithms that involve simulating multiple reflections of radiation such as daylight simulation or image rendering in general [6–10].

Over the years, different approaches have been developed to tackle this problem using different kinds of numerical algorithms. The multiple reflections of shortwave radiation can, for example, be handled by tracing the reflected photons using a Monte Carlo approach [11–15]. For the analysis of more mono-directed rays originating from discrete light or radiation sources, raytracing algorithms can be used to follow the radiation on its path through the modeled environment, similar to the algorithms implemented in the microclimate model ENVI-met to calculate the shadow casting [16–19]. Another group of algorithms approaches the problem from the other side, by not tracing the radiation itself, but analyzing the view relations between the different emitting and receiving elements in a scene or environment. These types of algorithms are generally categorized as radiosity approaches, with several sub-groups depending on the way the individual interactions between the elements are solved (e.g., matrix radiosity, progressive radiosity, or wavelet radiosity) [20–27].

While these methods give good results in smaller environments such as indoor spaces, they are hardly applicable for outdoor situations where an enormous number of elements and view relations between points in space need to be analyzed due to the larger and more complex geometries. Moreover, different to indoor environments, where the radiative fluxes are often calculated only once for a given setting, the situation in an outdoor environment is constantly changing due to movement of the sun and the resulting heating and cooling of surfaces. Hence, the radiative situation needs to be solved frequently—similar to the shadow casting, which is typically updated every few minutes [28].

Modern microclimate models such as the CFD-model ENVI-met that considers turbulence and is based on solving Reynolds–Navier–Stokes equations [28–31], allow for a very precise representation of the urban environment: Besides the detailed plant model [32,33], ENVI-met recently introduced an Accurate In-Canopy Radiation Transfer scheme, which accounts for scattering and attenuation of shortwave radiation within trees' canopies [34]. Buildings cannot only be digitized with a variety of different materials but also individual façades of the same building can consist of different materials [28,35,35,36]. The same is possible with the ground surface, where different surface types can be assigned to every grid. Since the different objects, materials, and surface types carry their own physical parameters such as albedo, emissivity, heat transfer coefficient, etc., the possibility to accurately replicate the variety of the urban environment drastically increases the accuracy of the model results. In previous versions of ENVI-met, however, the effects of different materials and surfaces were only calculated explicitly for primary radiation [37]. The distribution of secondary radiative fluxes (reflected shortwave radiation and longwave radiation emitted from objects) was, due to lack of memory and to save computational effort, carried out using a simplification, where instead of the actual reflected shortwave radiation and longwave radiation emitted, averages over all façades and surfaces within the model area were used [37]. This simplification could in some instances lead to a rather low accuracy that has been reported in evaluation studies comparing modeled against measured values of radiation and mean radiant temperature (MRT) [30,38–44].

In this paper, a newly developed method is introduced to overcome these shortcomings. It enables the simulation of fluxes of reflected shortwave and emitted longwave radiation within the urban environment with a much higher level of detail, considering the actual objects, i.e., façades, surfaces, and trees "seen" by a grid cell. By simulating secondary radiation in greater detail, differences in local microclimates can be identified more clearly and influences of, e.g., highly reflective surfaces or vegetation cover onto bioclimate indices such as physiological equivalent temperature (PET) can be taken into account. In a large ENVI-met proof-of-concept simulation featuring a multitude of different surfaces and plants, the new method is compared against the previous approach. The comparison should evaluate the accuracy of the new radiation scheme and examine the impacts of the more accurate secondary radiation modeling on the simulated local microclimate.

2. Model Description

In the following, the old averaged view factor concept used in ENVI-met is described (Section 2.1) and the advancements of the new indexed view sphere (IVS) algorithm are laid out in detail (Section 2.2). To evaluate the advantages of the new IVS module, a proof-of-concept simulation is conducted comparing the results of both concepts (Section 2.3).

2.1. The Averaged View Factor (AVF) Concept

To describe the radiative situation and solve the interactions between different elements, ENVI-met version 4.4.5 and prior used a generalized visibility concept based on averaged view factors (AVF). In the AVF concept, first, a three-dimensional ray tracing analysis is performed for every cell. Starting from a grid cell's center, rays are being shot for every 10° height and 10° azimuth angle creating a sphere consisting out of 18 × 36, thus 648 individual view facets. While calculating these 648 individual view facets for every grid cell can be very time consuming, it only needs to be performed once for a model simulation as the objects seen by a cell do not change over the course of a simulation. Based on the object type (sky, building, plant, ground surface) seen by the view facet/hit by the ray and the total number of view facets, averaged view factors of sky σ_{Sky}, vegetation σ_{Veg}, buildings σ_{Bldg}, and ground surfaces σ_{Grnd} were stored as single values for each grid cell. By only saving the view factors for the different object types instead of the actual façade elements, ground surfaces, and plant sections seen in the facets, a lot of memory can be saved. However, since no information is stored about which individual elements are in radiative exchange with the cell, the calculation of secondary radiative fluxes cannot take into account the radiation exchanges between actual objects seen but rather has to resort to an approximation. Instead of individual information about radiation received from particular objects, the secondary radiation is approximated by combining a grid cells' view factor for buildings, plants, and ground surfaces with averaged values of reflected and emitted longwave radiation for all buildings, plants, and ground surfaces over the entire model domain [19,28,37].

The received secondary radiation of a cell—in this case, the received shortwave reflected radiation from buildings—is, thus, calculated by:

$$Q_{swrefl,in}(i,j,k) = \sigma_{Bldg}(i,j,k) \cdot Q_{Bldg,swrefl} \tag{1}$$

with $\sigma_{Bldg}(x,y,z)$ as the building view factor (0 to 1) of cell i,j,k and $Q_{Bldg,swrefl}$ as the averaged reflected shortwave radiation from all buildings calculated by:

$$Q_{Bldg,swrefl} = \frac{1}{N}\sum_{a=1}^{N} \propto (a) \cdot Q_{sw}(a) \tag{2}$$

with N being the total number of façades in the model, $\propto$ the albedo of façade a, and Q_{sw} the shortwave radiation in front of façade a. The same calculations are carried out taking into account averaged surface temperature values of buildings, leaves, or ground surfaces for the estimation of received secondary longwave radiation.

To account for changes in the radiative situation, these calculations (Equations (1) and (2)) and their respective counterparts for other object types and emitted longwave radiation need to be solved regularly, i.e., every 15 min by default in ENVI-met.

The consequence of this simplification is that every point with an identical view factor for buildings, ground surfaces, or vegetation will receive the exact same amount of reflected shortwave and emitted longwave radiation.

While this simplification does save memory, it also leads to very unrealistic results: Given the same view factors, a cell would receive the exact same amount of reflected shortwave and emitted longwave radiation independent of the actual radiative processes in its vicinity. This could lead to identical reflected shortwave fluxes in front of north and south façades or high-reflective façade materials or low-reflective façade materials.

2.2. The Indexed View Sphere (IVS) Algorithm

To overcome this simplification, the indexed view sphere (IVS) algorithm was developed and implemented into ENVI-met. The main idea behind IVS is to save the results (i.e., the objects seen by a cell in a certain view angle) of the initial geometrical analysis so that it is possible to relate the calculated view factors back to the contributing elements of the urban scene. Its fundamentals are comparable to sky view factor calculations based on fish-eye imagery where the amount of sky pixels per annulus ring is analyzed (Figure 1a) [45–48]. To reduce the memory needed, the new IVS does not calculate the same amount of view facets in all height angles of the view sphere (Figure 1b).

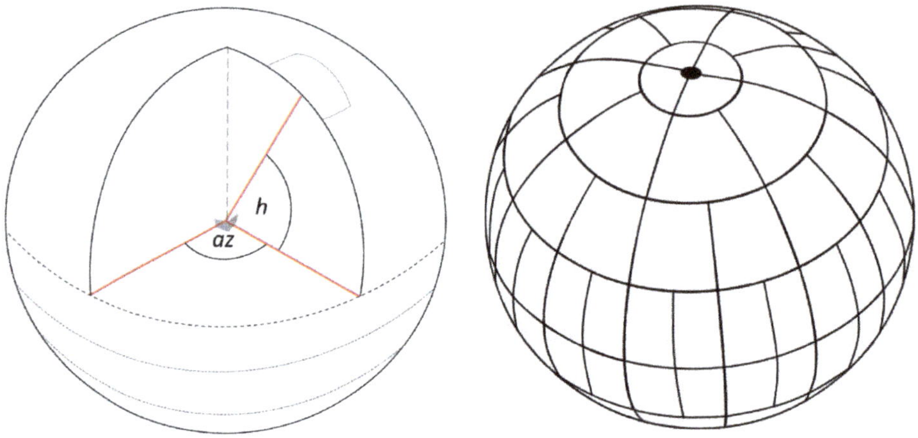

(a) (b)

Figure 1. View sphere for an exemplary grid cell to be analyzed depicting azimuthal angle az and height angle h definition (**a**) and individual view facet distribution per height angle (**b**) showing that view facet count decreases with increased height angles.

Since the resulting surface area for a given azimuthal angle decreases with increased height angles, the size of the azimuthal angle and, thus, of the view facets can be set larger with little to no information loss. The size of the azimuthal angles and, thus, the view facets is calculated by a function of the height angle. The user only defines the resolution of the height angles and the azimuthal angles at the equator of the cell. The number of azimuthal angles, i.e., the number of distinct view facets for a given height angle is then calculated by:

$$\lambda(h) = \frac{360}{az_{eq}} \cdot \cos(h) \tag{3}$$

with az_{eq} as the user defined size of azimuthal angle at the sphere's equator and the height angle h. To ensure all cardinal directions are represented, a minimum of four azimuthal angles is set for height angles less than $|90°|$. For height angles of $\pm 90°$ only one view facet is obtained, since azimuthal angles would be indifferent for these height angles.

By reducing the number of view facets per cell, an immense amount of memory can be saved without losing too much information about objects seen by a grid cell. Where a 10° height and azimuthal angle would previously result in 648 view facets, the same height and azimuthal resolution at the equator now results in 414 view facets. In combination with more efficient data structures, this decreases memory demand by around 40% and enables the possibility to store pointers to the objects seen by cells in a particular direction. By storing a direct link to objects seen by a cell, the actual radiation reflected/emitted by these objects can now be used to calculate a cell's received reflected shortwave and longwave radiation instead of using averaged values. Taking into account this radiation received, the cell alters its own reflection of shortwave and emission of longwave radiation and in turn contributes to the radiation received of other cells.

To calculate the received reflected shortwave and emission of longwave radiation coming from buildings or ground surfaces, the individual contribution of the objects is weighted by the view angle of the facet and simply added up:

$$Q_{Build,Grnd,in}(i,j,k) = \sum_{a=1}^{F} \omega(a) \cdot \tau_{Veg}(a) \cdot Q_{Build,Grnd,out} \quad (4)$$

with F being the total number of facets of the view sphere, $\omega(a)$ as the weight of the view facet a, $\tau_{Veg}(a)$ as the transmission factor accounting for reduction of the visibility due to vegetation between the cell i, j, k and the objects seen, and $Q_{Build,Grnd,out}$ as the outgoing reflected shortwave or emitted longwave radiation by the seen objects.

Secondary radiation emitted by vegetation is accounted for by inverting the transmission factor:

$$Q_{Veg,in}(i,j,k) = \sum_{a=1}^{F} \omega(a) \cdot (1 - \tau_{Veg}(a)) \cdot Q_{Veg,out} \quad (5)$$

with $Q_{Veg,out}$ as the outgoing reflected shortwave or emitted longwave radiation by the seen plants.

While the height angles are constant within a view sphere, the azimuthal angle and, thus, the number of view facets changes with increased height angles. This implies that the weighing factor is not identical for all view facets, but depended on the number of view facets in a particular height ring. The individual weighting factor accounting for the contribution of a view facet is calculated by:

$$\omega = \frac{1}{\lambda(h)} \cdot \frac{1}{r_{cnt}} \quad (6)$$

with $\lambda(h)$ as the number of view facets in a given height angle and r_{cnt} as the number of height rings for a particular view direction, e.g., downward view or upward view.

The transmission factor for vegetation τ_{Veg} is calculated within the raytracing and accounts for a partial obstruction of radiation due to vegetation in the ray's path. The calculation of transmission is carried out using the exponential extinction coefficient accounting for leaf orientation φ (currently set to 0.5), the local leaf area density (LAD), and the path length through the vegetation cells:

$$\tau_{Veg} = \mathrm{e}^{-(\varphi \cdot \, \mathrm{LAD} \cdot dRay)} \quad (7)$$

Further advancements have been undertaken with regards to the ray tracing algorithm. While in previous versions, the segment length for the ray would be determined only once for a given angle, the new algorithm tries to find an appropriate length for the ray segments based on current grid dimensions as well as the azimuthal and height angle. Determining a segment's length is not only critical for the calculation speed as short rays drastically increase computational effort but also for the quality of the resulting radiation calculation. With the discretization of space in models such as ENVI-met an optimal ray length is very hard to determine. With typical cell dimensions of 2 to 5 m in x, y, and z [28], too short ray segments might lead to an over-representation of cells, which are only marginally clipped by a ray, since the whole cell would count as "hit" by the ray. Too large ray segments, however, might lead to not detecting/not taking into account objects in the path of a ray.

To determine an optimal vector length increment for the ray trace, given the azimuthal and height angle, the normalized contribution weight (dx_{nwght}, dy_{nwght}, and dz_{nwght}) of the x, y, and z axis onto the resulting vector is calculated at first:

$$\begin{aligned} \begin{pmatrix} dx_{wght} \\ dy_{wght} \\ dz_{wght} \end{pmatrix} &= \begin{pmatrix} |\cos(h)\cdot \cos(az)| \\ |\cos(h)\cdot \sin(az)| \\ |\sin(h)| \end{pmatrix} \\ dxyz_{Sum} &= dx_{wght} + dy_{wght} + dz_{wght} \\ \begin{pmatrix} dx_{nwght} \\ dy_{nwght} \\ dz_{nwght} \end{pmatrix} &= \begin{pmatrix} dx_{wght}/dxyz_{Sum} \\ dy_{wght}/dxyz_{Sum} \\ dz_{wght}/dxyz_{Sum} \end{pmatrix} \end{aligned} \tag{8}$$

By taking into account the cell dimensions of the current cell, the normalized weights for all axes are then used to scale the length of the three-dimensional vector:

$$dRay(i,j,k) = \vartheta\cdot\left(dx(i,j,k)\cdot dx_{nwght} + dy(i,j,k)\cdot dy_{nwght} + dz(i,j,k)\cdot dz_{nwght}\right) \tag{9}$$

with ϑ as a scaling factor of the vector length. This scaling factor should ensure that cells are detected, which only partially lie along the path of the vector, i.e., a segment of the ray ends within the cell's boundaries. While a small scaling factor ensures a very precise ray tracing detecting all cells hit by a ray, it also leads to increased calculation time. Even more importantly, cells that might only lie very marginally in the path of the ray, i.e., clipped only very slightly, may be overrepresented in the further calculation. Test simulations showed that a scaling factors between 0.25 and 0.5 led to a good compromise between accuracy of the ray tracing and representation of cells clipped by the ray.

To determine the increments in the x, y, and z direction, the calculated vector length $dRay(i,j,k)$ is then again calculated by:

$$\begin{pmatrix} dx_{seg} \\ dy_{seg} \\ dz_{seg} \end{pmatrix} = \begin{pmatrix} dRay(i,j,k)\cdot \cos(h)\cdot \cos(az) \\ dRay(i,j,k)\cdot \cos(h)\cdot \sin(az) \\ dRay(i,j,k)\cdot \sin(h) \end{pmatrix} \tag{10}$$

After each iteration, the ray length for the current grid cell is updated to account for non-equidistant gridding in the model.

The new IVS calculation method for calculating secondary radiation transfers can be enabled in the simulation settings of the SIMX file. By default, the module is switched off. If enabled, the user is able to adjust the height angles and azimuthal angles at the sphere's equator. To save memory, the simulation settings allow us to define a high-resolution and a low-resolution angle pair. After a cut-off height defined, the model switches from the sphere definition of the high-resolution angles to the low-resolution angles. Similar to ENVI-met's vertical splitting (see [28]), radiation processes near the surface can, thus, be modeled with higher accuracy while processes of secondary radiation transfer above are carried out in a coarser resolution. This not only saves memory but also computational power. In case the user does not want to use this option, the high-resolution and low-resolution angles can simply be entered identically.

2.3. Proof-of-Concept Simulation

The advancements of ENVI-met's new IVS module are evaluated by comparing the model results of a simulation featuring the old AVF concept against a simulation featuring the new IVS algorithm, respectively. The model results will first be compared against each other with regard to longwave and shortwave radiation patterns, MRT, and potential air temperatures for different heights and times during the day. Furthermore, the impact of simulating secondary radiative transfers using IVS is examined by comparing the thermal comfort index PET between the two simulations.

In order to find the expected large differences in radiation budgets—and, thus, in the resulting microclimate conditions—between varying surface and building materials, the modeled area was chosen based on high spatial variability. The model area is located at the edge of Central Park in New York City, NY, USA featuring an additional high recognition value because of its iconic urban morphology surrounding the Columbus Circle. Varying building heights, differing materials such as glass, concrete, or brick and specific roof types such as roofs featuring greenery or cool (high-albedo) materials, which are common heat mitigation measures [49,50], already show the heterogeneity of the model area. Furthermore, the great differences in surface materials (meadows, rocks, sand pitches, streets, and pavements) together with the dense tree vegetation in Central Park located in a model area's lower right will enable us to show the complexity in secondary radiative fluxes that can be modeled by the new IVS module. To show the large variety of building and surface materials, the different materials have been visualized by specific colors (Figure 2). Table 1 gives an overview on the parameters of these materials that are based on the Midrise Apartment Post 1980 Standard for American Housing in NYC [51]. Building footprints and street tree locations were retrieved from the NYC open data portal [52]. Surface and building materials as well as park trees were digitized based on aerial imagery.

Figure 2. Example model area at the edge of Central Park in NYC modeled for the proof-of-concept simulations. Different colors indicate different material properties. Non-default ENVI-met database item properties are defined in Table 1. Grey façades/roofs indicate concrete walls and default roofs, respectively; blue transmissive façades indicate glass walls; red façades indicate brick walls; white roofs indicate cool roofs; green façades indicate default roofs with default greenery 01NASS being applied; green surfaces indicate default open soil 0100SL and default grass 0100XX; black surfaces indicate default asphalt road 0100ST; grey surfaces indicate default pavement 0100PP; pink surfaces indicate basalt rock profile entirely filled with default 0000BA soil; yellow surfaces indicate default sand profile 0100SD.

The meteorological boundary conditions were taken from an Energy Plus Weather File of NYC [53]. To resemble a hot summer day with clear sky conditions, July 21st was selected as simulation date. The simulation has been run for 24 h. Figure 3 shows the meteorological conditions for the whole simulation period provided by the EPW-file.

Table 1. Summary of material properties in the proof-of-concept simulation [51].

	Layer	Name	Albedo	Absorptivity	Emissivity	Thickness [m]	Conductivity [W (m K)$^{-1}$]	Density [kg m^{-3}]	Specific Heat (J (kg K)$^{-1}$)
Concrete Wall	Outside	Concrete	0.2	0.8	0.9	0.2	1.311	2240	836.8
	Middle	Steel frame Insulation	0.2	0.8	0.9	0.08	0.049	265	837
	Inside	1/2IN Gypsum	0.2	0.8	0.9	0.01	0.16	785	830
Glass Wall	All	Glass	0.25	0.05	0.94	3 × 0.01	0.029	830	840
Brick Wall	All	Brick	0.15	0.85	0.9	3 × 0.08	0.66	1500	1000
Default Roof	Outside	Default membrane	0.2	0.8	0.9	0.01	0.16	1121	1460
	Middle	Insulation	0.2	0.8	0.9	0.14	0.049	265	837
	Inside	Metal Decking	0.2	0.8	0.9	0.01	45	7680	418
Cool Roof	Outside	Cool material	0.65	0.35	0.9	0.02	0.85	1200	1200
	Middle	Insulation	0.2	0.8	0.9	0.14	0.049	265	837
	Inside	Metal Decking	0.2	0.8	0.9	0.01	45	7680	418

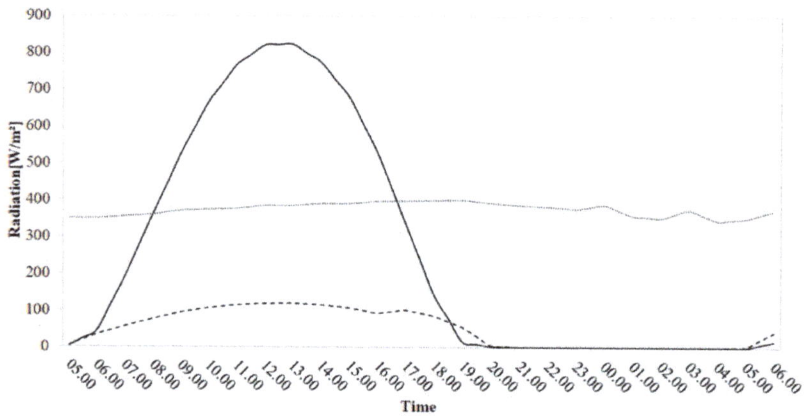

(a)

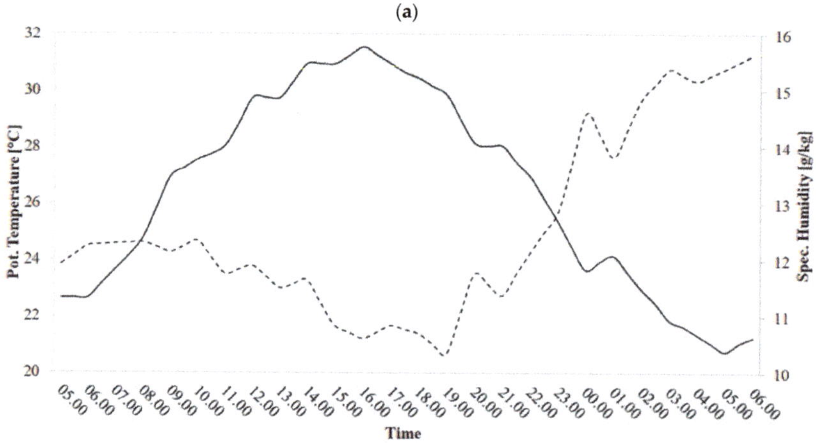

(b)

Figure 3. Meteorological boundary conditions for the proof-of-concept simulation depicting direct shortwave (black line), diffuse shortwave (dashed line), and longwave radiation (dotted line) (**a**) as well as potential air temperature (black line) and specific air humidity (dashed line) (**b**).

General simulation and model area properties are described in Table 2. Air temperature within buildings is held constant at 20 °C to account for the strong air conditioning cooling used in almost every building in NYC. Radiation scheme settings for the IVS simulations feature medium resolutions for lower cells at pedestrian level and low resolutions for higher cells above 8 m. Due to the large size of the model area featuring high-rise buildings and covering multiple different urban morphologies and typologies in a high spatial resolution, IVS settings could not feature higher view facet resolutions due to memory limitations. With these settings, the simulation yields around 50 GB of RAM for the IVS simulation.

Table 2. Parameters of the proof-of-concept simulations.

General Properties of Both Simulations	
Start Date and Time (Local)	21.07.2017 05:00
Duration [h]	24
Wind Speed [m s^{-1}]	2
Wind Direction [°]	325
Meteorological Boundary Conditions	Full Forcing
Location Lat (Lower Left Corner)	40.76° N
Location Lon (Lower Left Corner)	−73.98° E
Dimensions	360 × 260 × 52
Resolutions (X, Y, Z) [m]	2.5 × 2.5 × 5
Lowest Grid Cell Split	Yes
Telescoping: Factor and Starting Height	30% above 180 m
Height of 3D Model Domain [m]	1434 m
Building Indoor Temperature Held Constant at	20 °C
View Factor Update Interval	30 days
Specific Radiation Scheme Settings for IVS Simulation	
Height Angle High Resolution Near Ground	15°
Azimuthal Angle High Resolution Near Ground	15°
Height Angle Low Resolution Above Height Boundary	30°
Azimuthal Angle Low Above Height Boundary	30°
Height Boundary between High and Low Resolution	8 m

3. Results and Discussion

To give an overview on the differences between AVF and IVS, visualizations of the case study simulation results are presented. Radiation patterns as well as temperature distributions and impacts on thermal comfort are compared using absolute value maps of both scenarios to evaluate the advancements provided by the new IVS scheme.

After the theoretical explanations in Section 2.2., principles in IVS modeling are visually demonstrated by the 3D view of the model area. Figure 4 shows the rays that were traced for one specific grid cell at 138, 150, 6 amid the center of Columbus Circle. Each ray represents a view sphere segment of the grid cell to be analyzed. It is emitted from the grid cell center and searches for objects in each view sphere direction. When a building or soil surface is hit, the raytracing is aborted, and the object information is stored. With increasing height angle and especially above the highest object height, the length of each raytracing segment is, to save computational time, drastically increased as the probability of hitting an object is diminishing.

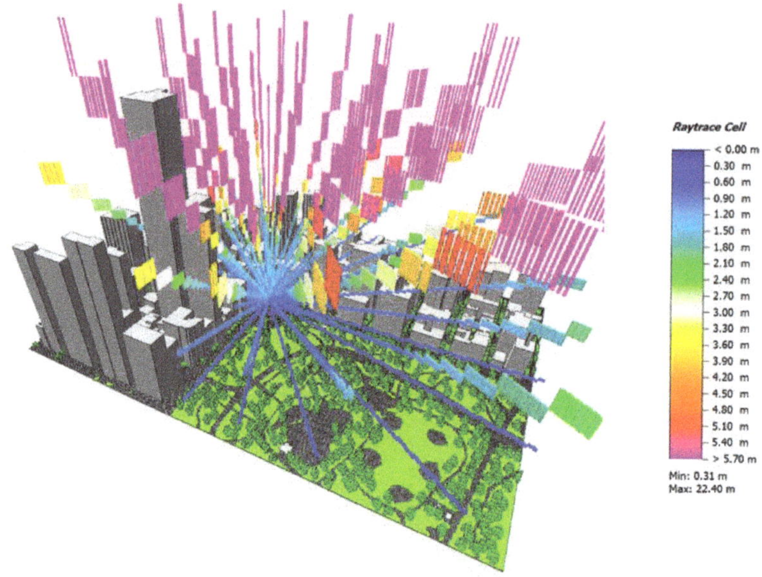

(**a**)

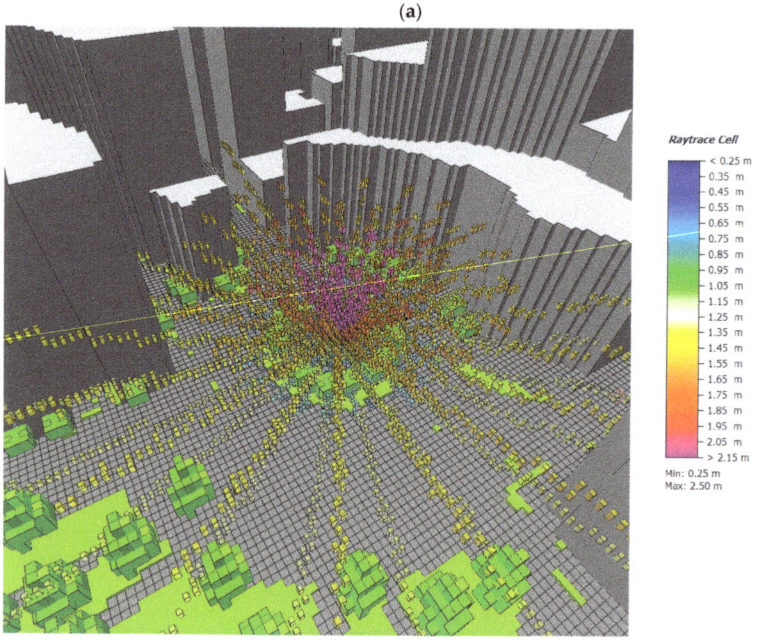

(**b**)

Figure 4. This specifically generated simulation output depicts the length of raytracing segments for one observed grid cell above the center of Columbus Circle. The 3D output view shows these raytracing lengths from a far perspective demonstrating the long ray distance and the length increase with height in (**a**) and from a nearby perspective with a 50 % decreased data cube size, 50 % data cube transparency, and data being filtered for height levels below k = 13 to visualize the high view sphere resolution and the resulting rays near the observed cell in (**b**).

ENVI-met provides multiple output variables to allow the analysis of radiation patterns. Variables containing shortwave or longwave radiation values are distinguished between the upper and lower hemispheres to allow for a specification from which direction the radiation is received. Analyzing, for example, longwave radiation distribution received from the lower hemisphere in a horizontal cut at 1.5 m height, most radiation input probably originates from soil surfaces, ground vegetation, and nearby façade elements. Figure 5a demonstrates that longwave radiation values cover a wider range and more distinct differences over the model at 13:00 in the IVS scenario compared against the AVF scenario. Especially the pattern of park streets can be recognized, as they emit more longwave radiation than their surroundings. That is caused by their low-albedo surfaces that heated up more during the morning hours in contrast to the street canyons, which have been shaded by buildings. In IVS, dotted patterns of lower longwave radiation values received by the lower hemisphere can be found in Central Park where trees cast shade and, thus, cool the ground below canopies leading to less longwave emission compared to the open surfaces in other park areas. These patterns are still visible in Figure 5b depicting the same results but for a horizontal cut in 82.5 m height. Furthermore, as the cut is performed in an intermediate height—above some low- and medium-rise buildings of the model area—differences in roof type effects can be examined. In IVS, the rather cold—and thereby not emitting much longwave radiation—cool and green roofs stand out against the traditional hotter low-albedo roofs that strongly contribute to the longwave radiation received in 82.5 m height. In AVF with its averaging approach, these different roof surface temperature differences while being simulated to not translate in differences of emitted longwave radiation as their contribution as specific object is not stored individually. Even differences in longwave radiation values caused by trees in Central Park are still recognized by IVS, even though spatial distances between the grid cells in 82.5 m height and the trees at ground level are quite large.

While surface temperature mainly drives the emitted longwave radiation patterns, reflected shortwave radiation is influenced by both albedo and shadowing. Figure 6a clearly demonstrates which areas are shaded during 13:00 when the sun's position is in the South—thus being at the model area's left due to its rotation. Figure 6a,b again show less pronounced differences in AVF. Especially the very small range between 90 and 144 W/m^2 at pedestrian level in AVF shows that most parts of the model area are rather homogenously affected by reflected shortwave radiation. The IVS results, however, accurately account for spatial relationships between objects and, thus, the models minimal reflected shortwave radiation of <5 W/m^2 for shaded areas but very large contributions of reflected radiation in high-albedo spots, for example above open sand at the ballpark in Central Park.

Furthermore, when performing a horizontal cut at 82.5 m height, we again find roofs in IVS to show higher values than other parts of the model area in contrast to AVF. However, their higher albedo, green and especially cool roofs now show higher values, as they reflect more shortwave radiation than the traditional low-albedo roofs. By examining these comparisons in Figures 5 and 6, it becomes clear that both longwave and shortwave contributions to a cell's radiation budget are underestimated by AVF. Urban heating caused by traditional building materials as well as cooling performances of cool or green roof application, hence, probably have been underestimated in previous ENVI-met modeling studies [30].

By using the model area average, AVF does not consider shortwave radiation reflections accurately in general, which is proved by two additional examples. It is firstly shown by the comparison of reflected shortwave radiation patterns for the horizontal cut at 82.5 m at 09:00 where the sun's position is in the east coming from the bottom side of the model area (Figure 7a). Since the circular shaped building at the Columbus Circle features glass façades and, thus, works like a magnifying glass, shortwave radiation is reflected from all sides to the center, which leads to a high accumulation in sum. By using IVS, these reflections are accounted for and values of up to 299 W/m^2 amid the circular-shaped building wall at the Columbus Circle are predicted by the model, whereas AVF only shows very low values around 25 W/m^2 here.

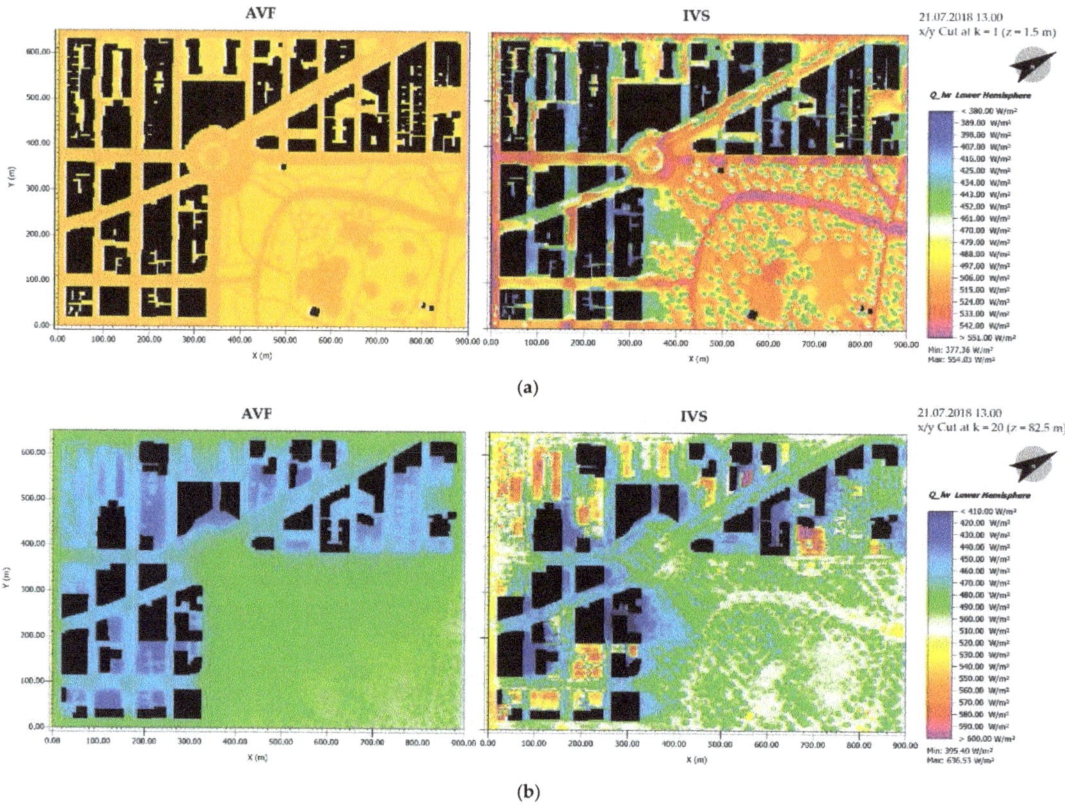

Figure 5. Comparison of proof-of-concept simulation results between AVF and IVS for longwave radiation received from the lower hemisphere at 13.00 in a height of 1.5 m (**a**) and in a height of 82.5 m (**b**), respectively.

Secondly, the pedestrian level horizontal cut of shortwave radiation reflections from the upper hemisphere also shows that, in contrast to AVF, IVS takes the reflection of glass façades into account (Figure 7b). As solar radiation at 13:00 originates from the south being at the left-hand side of the model area, shortwave reflected radiation coming from upper wall parts is modeled to be received in urban canyons at pedestrian level on the left-hand side of glass-façade buildings. The total amount, however, is comparatively low, as solar radiation is mostly reflected upwards from soil or roof surfaces and is, thus, included in the lower hemisphere values. Besides the primary reflection from wall surfaces, secondary reflected radiation can be found in Central Park, where primary reflected solar radiation from soil surfaces and grass is secondarily reflected downwards by tree canopies.

In contrast to the massively differing radiation patterns, at first glance, we only find small discrepancies in potential air temperature results for a horizontal cut at pedestrian level at 13:00 between AVF and IVS (Figure 8a). In general, IVS features slightly lower values of around 0.2–0.4 K. However, we find distinct differences between AVF and IVS in parts where high discrepancies of secondary radiation fluxes were observed. Especially the cooler regions in the southern/left part of the model area are more pronounced in IVS, as they receive less emitted longwave and less reflected shortwave radiation due to the shadows of larger skyscrapers next to them.

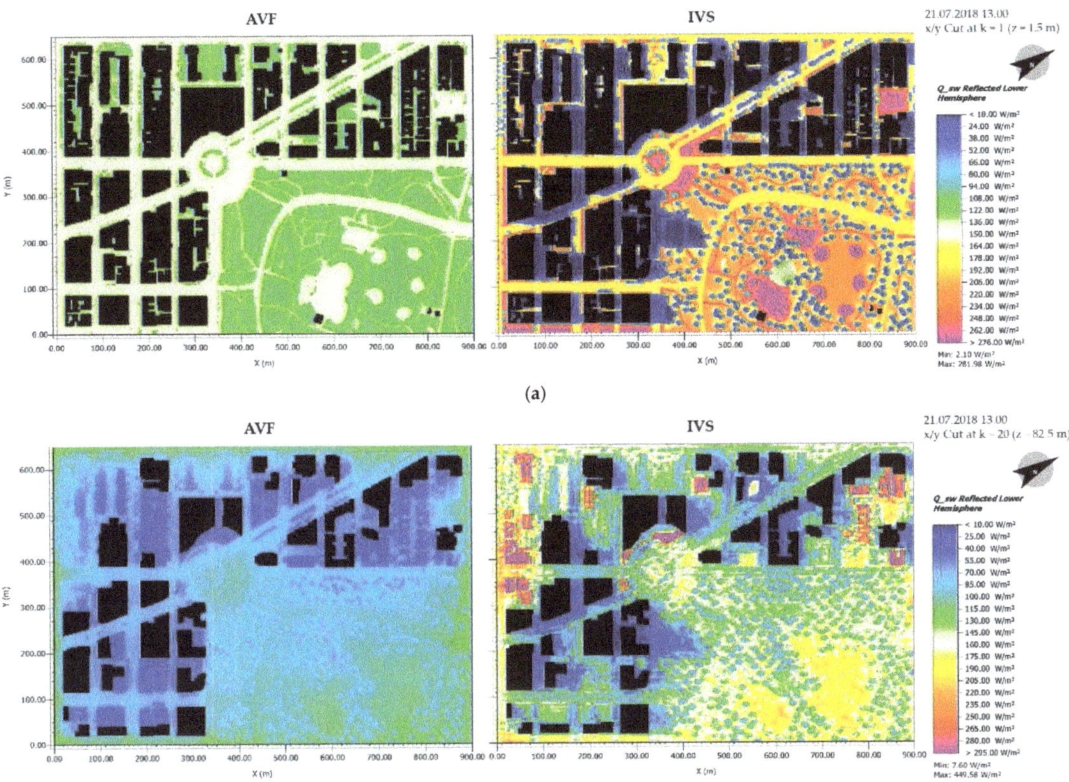

Figure 6. Comparison of proof-of-concept simulation results between AVF and IVS for reflected shortwave radiation received from the lower hemisphere at 13.00 in a height of 1.5 m (**a**) and in a height of 82.5 m (**b**), respectively.

Analysis of MRT corroborates that by showing even larger deviations between AVF and IVS for the same height and timestep (Figure 8b). According to the radiation patterns the MRT is based on, MRT also features a wider range of values in IVS.

To address the magnitude of urban heat stress and the effects of mitigation strategies properly, thermal comforts indices are often used as holistic indicators in urban environments. These indices take several parameters such as air temperature, humidity, wind speed, and MRT into account [54,55]. In this study, the thermal comfort index PET is used representing a commonly used index for outdoor environments of temperate climate zones [56–58]. When analyzing PET in Figure 8c, we find that major patterns are the same between AVF and IVS. However, the range is again wider and more differentiated in IVS simulation. As discrepancies in wind field and humidity are assumed to be marginal by the use of different radiation schemes and as temperatures are shown to be only slightly different in Figure 8a, deviations are mainly based on the differences in MRT patterns, which are in turn a key factor for thermal comfort and, thus, PET calculation.

These findings agree with the analysis of nighttime results shown in Figure 9. Longwave radiation emissions are based on the object temperatures. During nighttime, these surface temperatures of soils, buildings, and plants are influenced by their thermal properties, which specify their behavior in storing and releasing heat, as well as the amount of solar radiation that has or has not been received during daytime. As both of these parameters vary massively over the model area, simulated surface temperatures and, hence, longwave radiation patterns should differ as such. However, only in IVS, these very fine and distinct patterns of emitted longwave radiation from the lower hemisphere are recognized for both urban canyons as well as Central Park in the horizontal cut in 1.5 m height at 03:00 (Figure 9a). AVF, in contrast, shows a uniform distribution of values around 415 W/m^2 for all parts of the model area. As thermal comfort indices take MRT into account and MRT is solely based on longwave radiation during nighttime, the high accuracy in modeling longwave radiation budgets, when using IVS compared to AVF, improves the accuracy and reliability of thermal comfort indices. High accuracy for modeling nighttime urban heat stress is especially needed for metropolitan regions where thermal stress at nighttime can lead to health implications such as insomnia or cardio-vascular diseases [59]. In the comparison of PET results at pedestrian level for 03:00 (Figure 9b), we indeed find some discrepancies in defining thermal hotspots between both scenarios. By including more precise MRT values into thermal comfort calculations, PET is refined and mostly decreased in almost all parts of the model area. Cooling influences of Central Park for the built-up areas, however, cannot be observed as the wind flows in from the top of the model area and leaves at the open park side at the bottom of the model area. Since the influence of both shortwave and longwave radiation patterns on thermal comfort proves to be very large, the new IVS scheme seems to be able to contribute to a more accurate analysis of urban thermal comfort in general.

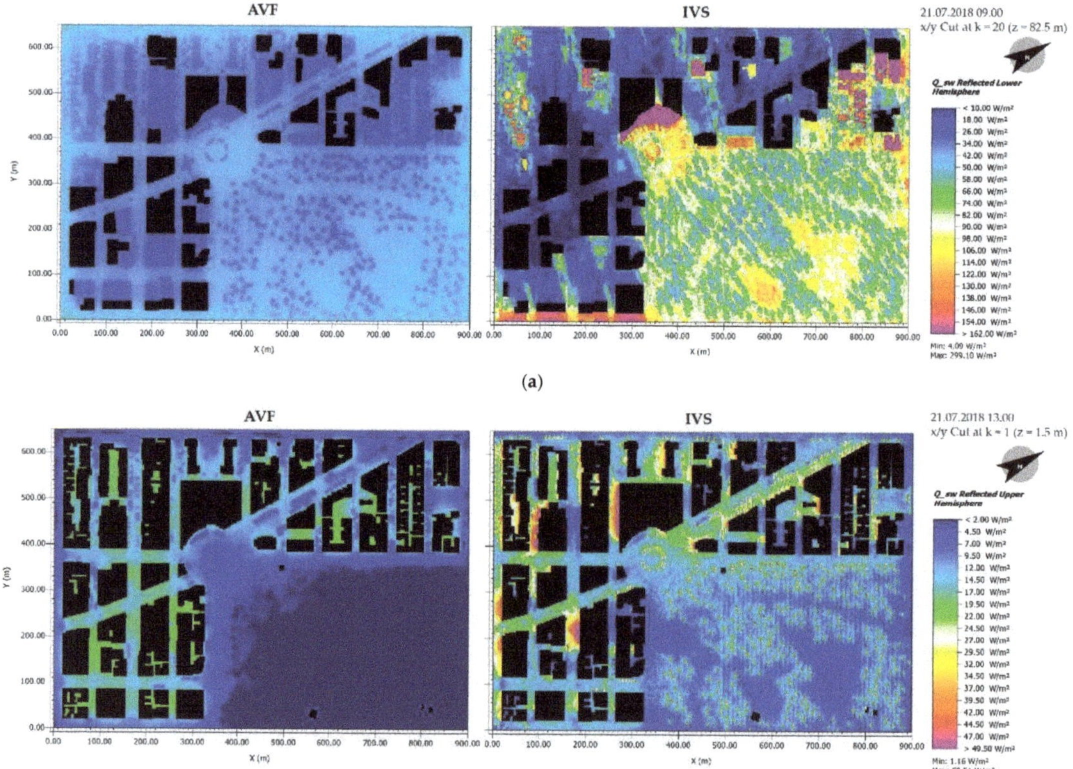

Figure 7. Comparison of proof-of-concept simulation results between AVF and IVS for reflected shortwave radiation received from the lower hemisphere at 09.00 in a height of 82.5 m (**a**) and for reflected shortwave radiation received from the upper hemisphere at 13.00 in a height of 1.5 m (**b**), respectively.

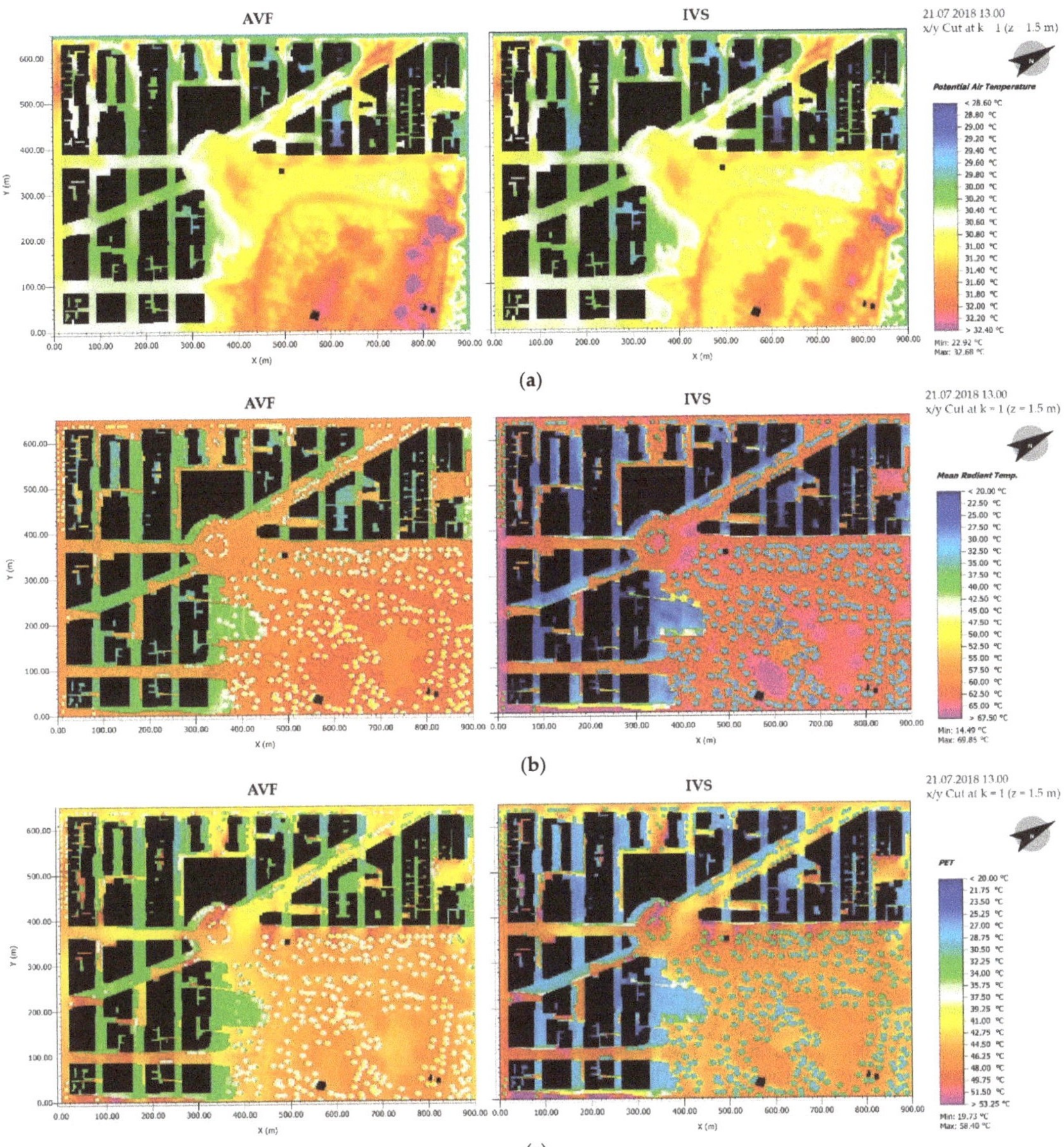

Figure 8. Comparison of proof-of-concept simulation results between AVF and IVS at 13.00 in a height of 1.5 m for potential air temperature (**a**), for mean radiant temperature (**b**), and for PET (**c**), respectively.

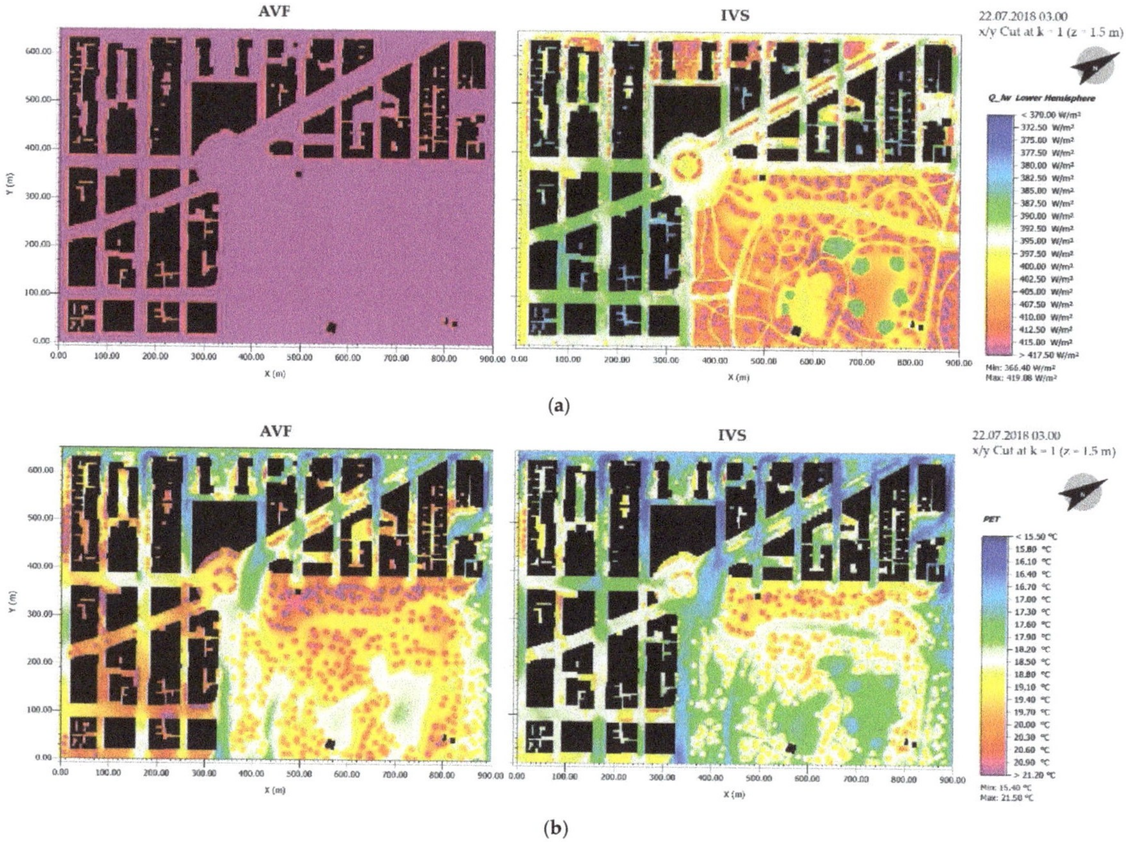

Figure 9. Comparison of proof-of-concept simulation results between AVF and IVS at 03.00 in a height of 1.5 m for longwave radiation received from the lower hemisphere (**a**) and for PET (**b**), respectively.

4. Conclusions

In order to improve microclimatological analysis of complex urban environments, a new radiation scheme is introduced into the microclimate model ENVI-met. In contrast to the old AVF scheme, where view factors are averaged over the whole model area, the new IVS scheme is able to analyze the surrounding objects of each grid cell and calculate their contribution to the local radiation budget. The high accuracy of modeling secondary radiative fluxes such as emitted longwave radiation and reflected shortwave radiation by using IVS was demonstrated by a proof-of-concept simulation. By taking into account shading of surfaces or a higher reflection by high-albedo materials, for example, fluxes of secondary shortwave and longwave radiation can be simulated in great detail. Several comparisons of different parameters, timesteps, and height levels between AVF and IVS showed that accurate surface properties and conditions affected the modeled microclimate conditions not only locally and for a specific timestep but also over larger distances and on the long-term when using IVS. Furthermore, the comparisons of PET indicate that a precise simulation of secondary radiation is very important to understand and prevent localized negative effects of, e.g., multiple shortwave reflections. These highly detailed results could be accomplished for a large, high-rise model area featuring yet rather low IVS resolutions. To gain an even higher accuracy in modeling results, the model could be simulated with an even higher IVS resolution and an even more accurate digitization of building surfaces and objects. However, the large demand in RAM limits the application potential for simulating larger model areas. In summary, the advantages provided by this new radiation scheme should, however, overturn this shortcoming. Future urban planning scenarios, architectural material questions, or heat mitigation measure studies using the microclimate model ENVI-met should, thus, be allowed to obtain more plausible and accurate results with the trade-off of a higher RAM demand.

Author Contributions: Conceptualization, H.S., T.S., and M.B.; methodology, H.S., M.B., and T.S.; software, H.S., T.S., and M.B.; validation, T.S.; formal analysis, T.S. and H.S.; investigation, T.S. and H.S.; resources, M.B.; data curation, T.S. and H.S.; writing—original draft preparation, T.S. and H.S.; writing—review and editing, H.S. and T.S.; visualization, T.S.; supervision, H.S., T.S., and M.B.; project administration, H.S. and T.S. All authors have read and agreed to the published version of the manuscript.

Funding: This research received no external funding.

Conflicts of Interest: The authors declare no conflict of interest.

References

1. Rosenfeld, A.H.; Akbari, H.; Bretz, S.; Fishman, B.L.; Kurn, D.M.; Sailor, D.; Taha, H. Mitigation of Urban Heat Islands: Materials, Utility Programs, Updates. *Energy Build.* **1995**, *22*, 255–265. [CrossRef]
2. Arnfield, A.J. An Approach to the Estimation of the Surface Radiative Properties and Radiation Budgets of Cities. *Phys. Geogr.* **1982**, *3*, 97–122. [CrossRef]
3. Arnfield, A.J. Street Design and Urban Canyon Solar Access. *Energy Build.* **1990**, *14*, 117–131. [CrossRef]
4. Oke, T.R. Urban Environments. In *The Surface Climates of Canada*; Bailey, W.G., Oke, T.R., Rouse, W.R., Eds.; McGill-Queen's University Press: Montreal, QC, Canada, 1997; pp. 303–327.
5. Santamouris, M. *Energy and Climate in the Urban Built Environment*; Routledge: Oxfordshire, UK, 2013; ISBN 9781134257904.
6. Cook, R.L.; Torrance, K.E. A Reflectance Model for Computer Graphics. *ACM SIGGRAPH Comput. Graph.* **1981**, *15*, 307–316. [CrossRef]
7. Mardaljevic, J. Daylight Simulation: Validation, Sky Models and Daylight Coefficients. Ph.D. Thesis, Loughborough University, Loughborough, UK, December 1999.
8. Hu, W.; Qin, K. Interactive Approximate Rendering of Reflections, Refractions, and Caustics. *IEEE Trans. Vis. Comput. Graph.* **2007**, *13*, 46–57. [CrossRef] [PubMed]
9. Brown, M.J.; Grimmond, S.; Ratti, C. *Comparison of Methodologies for Computing Sky View Factor in Urban Environments*; Los Alamos National Lab. (LANL): Los Alamos, NM, USA, 2001.
10. Hämmerle, M.; Gál, T.; Unger, J.; Matzarakis, A. Comparison of Models Calculating the Sky View Factor Used for Urban Climate Investigations. *Theor. Appl. Climatol.* **2011**, *105*, 521–527. [CrossRef]
11. Howell, J.R. Application of Monte Carlo to Heat Transfer Problems. In *Advances in Heat Transfer*; Irvine, T.F., Hartnett, J.P., Eds.; Elsevier: Amsterdam, The Netherlands, 1969; Volume 5, pp. 1–54.
12. Howell, J.R. The Monte Carlo Method in Radiative Heat Transfer. *J. Heat Transf.* **1998**, *120*, 547–560. [CrossRef]
13. Wang, Z.-H. Monte Carlo Simulations of Radiative Heat Exchange in a Street Canyon with Trees. *Sol. Energy* **2014**, *110*, 704–713. [CrossRef]
14. Aguerre, J.P.; Fernández, E.; Besuievsky, G.; Beckers, B. Computing Urban Radiation: A Sparse Matrix Approach. *Graph. Models* **2017**, *91*, 1–11. [CrossRef]
15. Lee, W.-L.; Liou, K.N.; Hall, A. Parameterization of Solar Fluxes over Mountain Surfaces for Application to Climate Models. *J. Geophys. Res. Atmos.* **2011**, *116*. [CrossRef]
16. Glassner, A.S. *An Introduction to Ray Tracing*; Morgan Kaufmann: Burlington, MA, USA, 1989; ISBN 9780122861604.
17. Shirley, P.; Morley, R.K. *Realistic Ray Tracing*, 2nd ed.; A K Peters, Ltd.: Natick, MA, USA, 2008; ISBN 9781568814612.
18. Meinzer, H.-P.; Meetz, K.; Scheppelmann, D.; Engelmann, U.; Baur, H.J. The Heidelberg Ray Tracing Model. *IEEE Comput. Graph. Appl.* **1991**, *11*, 34–43. [CrossRef]
19. Bruse, M. ENVI-Met 3.0: Updated Model Overview. 2004. Available online: https://www.envi-met.com/ (accessed on 11 April 2021).
20. Chelle, M.; Andrieu, B. The Nested Radiosity Model for the Distribution of Light within Plant Canopies. *Ecol. Model.* **1998**, *111*, 75–91. [CrossRef]
21. Wallace, J.R.; Elmquist, K.A.; Haines, E.A. A Ray Tracing Algorithm for Progressive Radiosity. In Proceedings of the 16th Annual Conference on Computer Graphics and Interactive Techniques, Boston, MA, USA, 1–4 August 1989; Association for Computing Machinery: New York, NY, USA, 1989; pp. 315–324.
22. Cohen, M.F.; Chen, S.E.; Wallace, J.R.; Greenberg, D.P. A Progressive Refinement Approach to Fast Radiosity Image Generation. In Proceedings of the 15th Annual Conference on Computer Graphics and Interactive Techniques, Atlanta, GA, USA, 1–5 August 1988; Association for Computing Machinery: New York, NY, USA, 1988; pp. 75–84.

23. Recker, R.J.; George, D.W.; Greenberg, D.P. Acceleration Techniques for Progressive Refinement Radiosity. In Proceedings of the 1990 Symposium on Interactive 3D graphics, Snowbird, UT, USA, 25–28 March 1990; Association for Computing Machinery: New York, NY, USA, 1990; pp. 59–66.
24. Helbig, N.; Löwe, H.; Lehning, M. Radiosity Approach for the Shortwave Surface Radiation Balance in Complex Terrain. *J. Atmos. Sci.* **2009**, *66*, 2900–2912. [CrossRef]
25. Fiala, P.; Kadlecová, E. Progressive Methods in the Numerical Modeling by the Finite Elements. In Proceedings of the 4th WSEAS International Conference on Applications of Electrical Engineering, Prague, Czech Republic, 12–14 March 2005.
26. Gortler, S.J.; Schröder, P.; Cohen, M.F.; Hanrahan, P. Wavelet Radiosity. In Proceedings of the 20th annual conference on Computer Graphics and Interactive Techniques—SIGGRAPH '93, Anaheim, CA, USA, 2–6 August 1993; ACM Press: New York, NY, USA, 1993; pp. 221–230.
27. Ayoub, M. A Review on Light Transport Algorithms and Simulation Tools to Model Daylighting inside Buildings. *Sol. Energy* **2020**, *198*, 623–642. [CrossRef]
28. Simon, H. Modeling Urban Microclimate: Development, Implementation and Evaluation of New and Improved Calculation Methods for the Urban Microclimate Model ENVI-Met. Ph.D. Thesis, Johannes Gutenberg-University, Mainz, Germany, 1 June 2016.
29. Bruse, M.; Fleer, H. Simulating Surface Plant Air Interactions inside Urban Environments with a Three Dimensional Numerical Model. *Environ. Model. Softw.* **1998**, *13*, 373–384. [CrossRef]
30. Tsoka, S.; Tsikaloudaki, A.; Theodosiou, T. Analyzing the ENVI-Met Microclimate Model's Performance and Assessing Cool Materials and Urban Vegetation Applications—A Review. *Sustain. Cities Soc.* **2018**, *43*, 55–76. [CrossRef]
31. Crank, P.J.; Sailor, D.J.; Ban-Weiss, G.; Taleghani, M. Evaluating the ENVI-Met Microscale Model for Suitability in Analysis of Targeted Urban Heat Mitigation Strategies. *Urban Clim.* **2018**, *26*, 188–197. [CrossRef]
32. Simon, H.; Lindén, J.; Hoffmann, D.; Braun, P.; Bruse, M.; Esper, J. Modeling Transpiration and Leaf Temperature of Urban Trees—A Case Study Evaluating the Microclimate Model ENVI-Met against Measurement Data. *Landsc. Urban Plan.* **2018**, *174*, 33–40. [CrossRef]
33. Shinzato, P.; Simon, H.; Duarte, D.H.S.; Bruse, M. Calibration Process and Parametrization of Tropical Plants Using ENVI-Met V4—Sao Paulo Case Study. *Archit. Sci. Rev.* **2019**, *62*, 112–125. [CrossRef]
34. Simon, H.; Sinsel, T.; Bruse, M. Introduction of Fractal-Based Tree Digitalization and Accurate In-Canopy Radiation Transfer Modelling to the Microclimate Model ENVI-Met. *Forests* **2020**, *11*, 869. [CrossRef]
35. Simon, H.; Kissel, L.; Bruse, M. Evaluation of ENVI-Met's Multiple-Node Model and Estimation of Indoor Climate. In Proceedings of the PLEA 2017—Passive and Low Energy Architecture, Edinburgh, UK, 3–5 July 2017; p. 9.
36. Forouzandeh, A. Prediction of Surface Temperature of Building Surrounding Envelopes Using Holistic Microclimate ENVI-Met Model. *Sustain. Cities Soc.* **2021**, *70*, 102878. [CrossRef]
37. Huttner, S. Further Development and Application of the 3D Microclimate Simulation ENVI-Met. Ph.D. Thesis, Johannes Gutenberg-University, Mainz, Germany, 4 May 2012.
38. Acero, J.A.; Herranz-Pascual, K. A Comparison of Thermal Comfort Conditions in Four Urban Spaces by Means of Measurements and Modelling Techniques. *Build. Environ.* **2015**, *93*, 245–257. [CrossRef]
39. Acero, J.A.; Arrizabalaga, J. Evaluating the Performance of ENVI-Met Model in Diurnal Cycles for Different Meteorological Conditions. *Theor. Appl. Climatol.* **2018**, *131*, 455–469. [CrossRef]
40. Salata, F.; Golasi, I.; de Lieto Vollaro, R.; de Lieto Vollaro, A. Urban Microclimate and Outdoor Thermal Comfort. A Proper Procedure to Fit ENVI-Met Simulation Outputs to Experimental Data. *Sustain. Cities Soc.* **2016**, *26*, 318–343. [CrossRef]
41. Sharmin, T.; Steemers, K.; Matzarakis, A. Microclimatic Modelling in Assessing the Impact of Urban Geometry on Urban Thermal Environment. *Sustain. Cities Soc.* **2017**, *34*, 293–308. [CrossRef]
42. Lee, H.; Mayer, H.; Chen, L. Contribution of Trees and Grasslands to the Mitigation of Human Heat Stress in a Residential District of Freiburg, Southwest Germany. *Landsc. Urban Plan.* **2016**, *148*, 37–50. [CrossRef]
43. Chen, L.; Ng, E. Simulation of the Effect of Downtown Greenery on Thermal Comfort in Subtropical Climate Using PET Index: A Case Study in Hong Kong. *Archit. Sci. Rev.* **2013**, *56*, 297–305. [CrossRef]
44. Roth, M.; Lim, V.H. Evaluation of Canopy-Layer Air and Mean Radiant Temperature Simulations by a Microclimate Model over a Tropical Residential Neighbourhood. *Build. Environ.* **2017**, *112*, 177–189. [CrossRef]
45. Steyn, D.G. The Calculation of View Factors from Fisheye-lens Photographs: Research Note. *Atmos. Ocean* **1980**, *18*, 254–258. [CrossRef]
46. Steyn, D.G.; Hay, J.E.; Watson, I.D.; Johnson, G.T. The Determination of Sky View-Factors in Urban Environments Using Video Imagery. *J. Atmos. Ocean. Technol.* **1986**, *3*, 759–764. [CrossRef]
47. Chapman, L.; Thornes, J.E. Real-Time Sky-View Factor Calculation and Approximation. *J. Atmos. Ocean. Technol.* **2004**, *21*, 730–741. [CrossRef]
48. Middel, A.; Lukasczyk, J.; Maciejewski, R.; Demuzere, M.; Roth, M. Sky View Factor Footprints for Urban Climate Modeling. *Urban Clim.* **2018**, *25*, 120–134. [CrossRef]
49. Santamouris, M.; Yun, G.Y. Recent Development and Research Priorities on Cool and Super Cool Materials to Mitigate Urban Heat Island. *Renew. Energy* **2020**, *161*, 792–807. [CrossRef]

50. He, Y.; Yu, H.; Ozaki, A.; Dong, N. Thermal and Energy Performance of Green Roof and Cool Roof: A Comparison Study in Shanghai Area. *J. Clean. Prod.* **2020**, *267*, 122205. [CrossRef]
51. Deru, M.; Field, K.; Studer, D.; Benne, K.; Griffith, B.; Torcellini, B.; Halverson, M.; Winiarski, D.; Liu, B.; Rosenberg, M.; et al. *U.S. Department of Energy Commercial Reference Building Models of the National Building Stock*; NREL/TP-5500-46861; National Renewable Energy Laboratory: Golden, CO, USA, 2010.
52. NYC Open Data, D. of I.T.& T. (DoITT). Building Footprints and Street Trees. Available online: https://data.cityofnewyork.us/Housing-Development/Building-Footprints/nqwf-w8eh (accessed on 4 January 2021).
53. Wilcox, S.; Marion, W. *Users Manual for TMY3 Data Sets*; National Renewable Energy Laboratory: Golden, CO, USA, 2008.
54. Höppe, P. The Physiological Equivalent Temperature. A Universal Index for the Biometeorological Assessment of the Thermal Environment. *Int. J. Biometeorol.* **1999**, *43*, 71–75. [CrossRef] [PubMed]
55. Matzarakis, A.; Mayer, H.; Iziomon, M.G. Applications of a Universal Thermal Index: Physiological Equivalent Temperature. *Int. J. Biometeorol.* **1999**, *43*, 76–84. [CrossRef]
56. Matzarakis, A.; Amelung, B. Physiological Equivalent Temperature as Indicator for Impacts of Climate Change on Thermal Comfort of Humans. In *Seasonal Forecasts, Climatic Change and Human Health: Health and Climate*; Thomson, M.C., Garcia-Herrera, R., Beniston, M., Eds.; Advances in Global Change Research; Springer: Dordrecht, The Netherlands, 2008; pp. 161–172. ISBN 978140206877-5.
57. He, X.; Miao, S.; Shen, S.; Li, J.; Zhang, B.; Zhang, Z.; Chen, X. Influence of Sky View Factor on Outdoor Thermal Environment and Physiological Equivalent Temperature. *Int. J. Biometeorol.* **2015**, *59*, 285–297. [CrossRef] [PubMed]
58. Gómez, F.; Cueva, A.P.; Valcuende, M.; Matzarakis, A. Research on Ecological Design to Enhance Comfort in Open Spaces of a City (Valencia, Spain). Utility of the Physiological Equivalent Temperature (PET). *Ecol. Eng.* **2013**, *57*, 27–39. [CrossRef]
59. Schinasi, L.H.; Benmarhnia, T.; De Roos, A.J. Modification of the Association between High Ambient Temperature and Health by Urban Microclimate Indicators: A Systematic Review and Meta-Analysis. *Environ. Res.* **2018**, *161*, 168–180. [CrossRef] [PubMed]

applied sciences

MDPI

Article

Determination of the Vertical Load on the Carrying Structure of a Flat Wagon with the 18–100 and Y25 Bogies

Oleksij Fomin [1], Alyona Lovska [2], Václav Píštěk [3] and Pavel Kučera [3,*]

[1] Department of Cars and Carriage Facilities, State University of Infrastructure and Technologies, Kyrylivska Str., 9, 04071 Kyiv, Ukraine; fomin_ov@gsuite.duit.edu.ua
[2] Department of Wagons, Ukrainian State University of Railway Transport, Feuerbach sq., 7, 61050 Kharkiv, Ukraine; alyonalovskaya@kart.edu.ua
[3] Institute of Automotive Engineering, Brno University of Technology, Technická 2896/2, 616 69 Brno, Czech Republic; pistek.v@fme.vutbr.cz
* Correspondence: kucera@fme.vutbr.cz; Tel.: +420-541-142-274

Featured Application: The data obtained during the study will be useful in the development and construction of innovative rolling stock. The results of the work will also contribute to increasing the efficiency of operation and safety of rolling stock by reducing their dynamic loading.

Abstract: The study deals with determination of the vertical load on the carrying structure of a flat wagon on the 18–100 and Y25 bogies using mathematic modelling. The study was made for an empty wagon passing over a joint irregularity. The authors calculated the carrying structure of a flat wagon with the designed parameters and the actual features recorded during field tests. The mathematical model was solved in MathCad software. The study found that application of the Y25 bogie for a flat wagon with the designed parameters can decrease the dynamic load by 41.1% in comparison to that with the 18–100 bogie. Therefore, application of the Y25 bogie under a flat wagon with the actual parameters allows decreasing the dynamic loading by 41.4% in comparison to that with the 18–100 bogie. The study also looks at the service life of the supporting structure of a flat wagon with the Y25 bogie, which can be more than twice as long as the 18–100 bogie. The research can be of interest for specialists concerned with improvements in the dynamic characteristics and the fatigue strength of freight cars, safe rail operation, freight security, and the results of the research can be used for development of innovative wagon structures.

Keywords: transport mechanics; flat wagon; carrying structure; dynamic load; dynamics modelling; service life

Citation: Fomin, O.; Lovska, A.; Píštěk, V.; Kučera, P. Determination of the Vertical Load on the Carrying Structure of a Flat Wagon with the 18–100 and Y25 Bogies. *Appl. Sci.* **2021**, *11*, 4130. https://doi.org/10.3390/app11094130

Academic Editor: María Isabel Lamas Galdo

Received: 4 April 2021
Accepted: 28 April 2021
Published: 30 April 2021

Publisher's Note: MDPI stays neutral with regard to jurisdictional claims in published maps and institutional affiliations.

Copyright: © 2021 by the authors. Licensee MDPI, Basel, Switzerland. This article is an open access article distributed under the terms and conditions of the Creative Commons Attribution (CC BY) license (https://creativecommons.org/licenses/by/4.0/).

1. Introduction

The purchase of innovative freight wagons requires large capital investments. A preliminary estimate has shown that renovating an existing rolling stock in terms of its efficient operation is more cost-effective than purchasing new transport means. Therefore, introduction of the measures aimed at decreasing loads on the railway vehicles under operational modes is of primary importance.

As known, the dynamic loading, conditioned by a lot of factors, impact mostly the strength and the operational life of the carrying structure of rail wagons and one of the most important factors is the joint irregularity. The periodic dynamic loading transferring from the bogie to the carrying structure results on fatigue behaviour. Therefore, for a purpose of a longer service life of a rail wagon, it is of importance to research into the dynamic loading on the carrying structure of a wagon for various types of bogie. The broad-gauge lines mostly use the 18–100 bogie which has been in operation since 1950s. This bogie has been modernized many times since then.

In addition, better dynamic characteristics of rail cars, provision of tensile strength and fatigue strength, longer service life, freight security during transportation can be achieved through the research into the optimal design of gear parts under the new rail wagons with a longer service life.

The narrow-gauge lines mostly use the Y25 bogie. This bogie has good operational characteristics and at present it has several modifications. Therefore, it is important to research the dynamic loading on the carrying structure of rail wagons with this bogie type and analyze a possibility to use it as an alternative to the 18–100 bogie.

2. Analysis of Recent Research and Publications

The mechanical resistance of a rail vehicle and measures for its improvements are studied in [1]. The research was made with the mathematic modelling. The results were used for grounding the application of a semi-active bogie suspension.

The dynamic analysis for a wagon with modified bogies is presented in [2]. The calculation was made for a Shimmns freight wagon in motion for the loaded and empty states. However, these studies do not include determination of the dynamic loading on a wagon with the actual parameters of the carrying elements and introduction of measures for a decrease in the dynamic loads.

The structural analysis of a modified freight wagon is given in [3]. The problem was solved by means of the finite element methods (FEM). The results of the calculation proved the efficiency of the solutions taken.

The modelling properties of a wagon multi-wheel system and its dynamic properties using computational modelling are described in [4]. This approach was made for a freight wagon on the Y25 bogie. However, the authors did not study the impact of the Y25 bogie on the dynamic load of the carrying structure.

The approach into an impact of a three-piece bogie suspension with two types of friction wedges on the vertical load is presented in [5]. The characteristics of dynamic response and comparison with various friction conditions for a friction wedge and the input frequencies were given in the study. The measures for improvements of the rail strength due to application of a new bogie are given in [6]. The authors also make suggestions regarding the freight bogie development. They substantiated the application of the Y25L in terms of the operational security of the rolling stock. However, it should be noted, that these studies do not include the impact of proposed solutions to the service life of rail vehicles.

Studies [7,8] present some measures to decrease the dynamic loading of the carrying structure of a rail wagon and to prolong the service life. The substantiation of the solutions was made by means of the mathematical modelling with the subsequent computer supporting. However, the authors do not explore a possibility to apply the bogie of the optimal spring suspension characteristics to decrease the dynamic loading of the carrying structure.

The literature analysis made it possible to conclude that the issue for determining the vertical load of the carrying structure of a wagon with the Y25 bogie, as an alternative variant to the 18–100 bogie, has not been studied yet. Thus, it requires appropriate research in the field.

3. The Objective and the Tasks of the Research

The objective of the article is to present special features for determining the vertical load of the carrying structure of a flat wagon with the Y25 bogie and substantiate the application of this bogie for broad-gauge lines as an alternative to the 18–100 bogie. To achieve the objective the following tasks were set:

- mathematical modelling of the dynamic loading on the carrying structure of a flat wagon with the design parameters with the 18–100 and Y25 bogies,
- mathematical modelling of the dynamic loading on the carrying structure of a flat wagon with the actual parameters with the 18–100 and Y25 bogies,
- determination of the design service life of the carrying structure of a flat wagon with the 18–100 and Y25 bogies.

4. The Presentation of the Main Content of the Study

The basic dynamic characteristics of the carrying structure of a wagon with the Y25 and 18–100 bogies were determined by means of the mathematical modelling. It was based on the mathematical model proposed by Professor Yu. V. Diomin and Associate Professor G. Yu. Cherniak [9]. The research was made in the plane coordinates. The study included the bouncing oscillations as one of the most frequent oscillations for rail vehicles in operation.

The calculation was made for a 13–401 flat wagon modernized for container transportation (see Figure 1). This wagon type was chosen as the prototype as it is the most widespread wagon for international transportation.

Figure 1. Flat car 13–401 (**a**) universal; (**b**) modified for container transportation.

The authors defined the dynamic loading of the carrying structure of a flat wagon with the design (initial) parameters on the 18–100 and Y25 bogies, and also a flat wagon with the actual parameters recorded during field tests.

The study included motion of an empty wagon, because it was the best way to study the dynamic loading when the wagon passed over a joint irregularity. The track was taken as a viscous elastic component. It was assumed that the track behaviour was proportional to both its deformation value and the speed of this deformation. Taking into account this assumption, the expression for determining the dissipative energy was obtained from the expression of the potential energy, by replacing the generalized coordinates in it with the generalized velocities and spring stiffnesses—the damping coefficients.

The dynamic computational model of the wagon is shown in Figure 2.

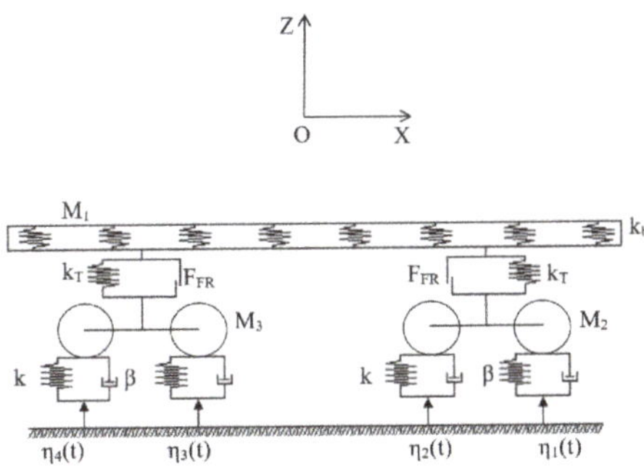

Figure 2. Computational model of the wagon.

The flat wagon was taken as a system of three solid bodies: frame and two bogies with suspension groups.

The following links for the system were assumed:

- displacements of the frame and the bogies of a flat wagon along the track axle were equal,
- wheelsets were moving without a sliding,
- due to absence of elastic elements in the axle–box suspension, the bouncing of the bogies was determined through the bouncing of the wheelsets.

The equations of motion for the designed model are as follows:

$$M_1\frac{d^2}{dt^2}q_1 + C_{1,1}q_1 + C_{1,3}q_3 + C_{1,5}q_5 = F_z, \tag{1}$$

$$M_2\frac{d^2}{dt^2}q_2 + C_{2,2}q_2 + C_{2,3}q_3 + C_{2,5}q_5 = F_\varphi, \tag{2}$$

$$M_3\frac{d^2}{dt^2}q_3 + C_{3,1}q_1 + C_{3,2}q_2 + C_{3,3}q_3 + B_{3,3}\frac{d}{dt}q_3 = F_{z,B1}, \tag{3}$$

$$M_4\frac{d^2}{dt^2}q_4 + C_{4,4}q_4 + B_{4,4}\frac{d}{dt}q_4 = F_{\varphi,B1}, \tag{4}$$

$$M_5\frac{d^2}{dt^2}q_5 + C_{5,1}q_1 + C_{5,2}q_2 + C_{5,5}q_5 + B_{5,5}\frac{d}{dt}q_5 = F_{z,B2}, \tag{5}$$

$$M_6\frac{d^2}{dt^2}q_6 + C_{6,6}q_6 + B_{6,6}\frac{d}{dt}q_6 = F_{\varphi,B2}, \tag{6}$$

$$F_z = -F_{FR}\left(\text{sign}\left(\frac{d}{dt}\delta_1\right) + \text{sign}\left(\frac{d}{dt}\delta_2\right)\right), \tag{7}$$

$$F_\varphi = F_{FR}l\left(\text{sign}\left(\frac{d}{dt}\delta_1\right) - \text{sign}\left(\frac{d}{dt}\delta_2\right)\right) \tag{8}$$

$$F_{z,B1} = F_{FR}\text{sign}\left(\frac{d}{dt}\delta_1\right) + k_1(\eta_1 + \eta_2) + \beta_1\left(\frac{d}{dt}\eta_1 + \frac{d}{dt}\eta_2\right), \tag{9}$$

$$F_{\varphi,B1} = -k_1a(\eta_1 - \eta_2) - \beta_1a\left(\frac{d}{dt}\eta_1 - \frac{d}{dt}\eta_2\right), \tag{10}$$

$$F_{z,B2} = F_{FR}\text{sign}\left(\frac{d}{dt}\delta_2\right) + k_1(\eta_3 + \eta_4) + \beta_1\left(\frac{d}{dt}\eta_3 + \frac{d}{dt}\eta_4\right), \tag{11}$$

$$F_{\varphi,B2} = -k_1a(\eta_3 - \eta_4) - \beta_1a\left(\frac{d}{dt}\eta_3 - \frac{d}{dt}\eta_4\right), \tag{12}$$

where M_1, M_2—mass and inertia moment of the carrying structure of the flat wagon at bouncing and galloping oscillations, respectively; M_3, M_4—mass and inertia moment of the first bogie facing the engine at bounding and galloping oscillations, respectively; M_5, M_6—mass and inertia moment of the second bogie facing the engine at bounding and galloping oscillations, respectively; C_{ij}—characteristics of elastic elements of the oscillating system; B_i—scattering function; a—half-base of a bogie; q_i—generalized coordinates corresponding to the advancing movement relative to the vertical axle and the angular displacement around the vertical axle; k_i—track stiffness; β_i—damping coefficient; F_{FR}—absolute friction force in a spring group.

The input parameters of the model were the technical characteristics of the carrying structure with the design and actual parameters, the bogies (Figures 3 and 4), and the disturbing force (Table 1).

Characteristics included in modelling
mass – 4.3 tons
moment of inertia – 3.0 ton·m^2
half-base – 0.925 m
spring suspension stiffness – 8,000 kN/m
specified friction coefficient – 0.1

Figure 3. 18–100 bogie and its main parameter.

Characteristics included in modelling
mass – 4.85 tons
moment of inertia – 6.7 ton·m^2
half-base – 0.9 m
spring suspension stiffness – 4,004.1 kN/m
specified friction coefficient – 0.1

Figure 4. Y25 bogie and its main parameter.

Table 1. Parameters of the disturbing force.

Track Parameter	Value
damping coefficient, kN·s/m	200
stiffness, kN/m	100,000
irregularity amplitude, m	0.01
irregularity length, m	25

The mass and the moment of inertia of the carrying structure of the flat wagon with the design parameters and with the actual parameters were determined through their spatial models in SolidWorks software [10–12] The mass of the carrying structure of the flat wagon with the design parameters was 15.6 tons, the moment of inertia 283.1 ton × m^2, and those with the actual parameters 11.1 ton and 102 ton × m^2, respectively (Figure 5).

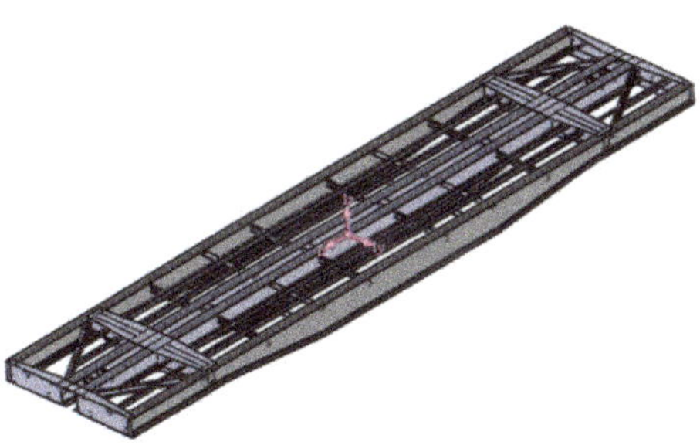

Parameter	Design	Actual
Mass, kg	11,520.34	11,100.85
Volume, m^3	1.47	1.42
Surface area, m^2	138.79	137.08
Moments of inertia, kg · m^2	Px = 5,771.33 Py=117,450.29 Pz=105,612.86	Px = 5,091.62 Py=116,294.06 Pz=102,108.23

Figure 5. Spatial model of the carrying structure of the flat wagon.

Differential Equations (1)–(6) were reduced to standard Cauchy problems and then they were integrated by the Runge–Kutta method [13–18]. The initial displacements and speeds were taken equal to zero. Based on the calculation, the basic parameters of dynamics for wagons with the 18–100 and Y25 bogie were found.

The accelerations on the carrying structure of the flat wagon with the actual parameters on the 18–100 and Y25 bogies in the center of gravity are given in Figures 6 and 7 and the accelerations in the areas of support on the bogies in Figures 8 and 9.

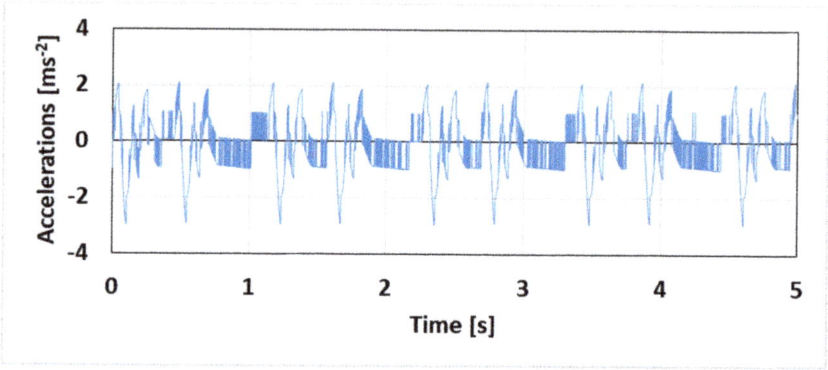

Figure 6. Accelerations of the flat wagon carrying structure in the center of gravity: bogie 18–100.

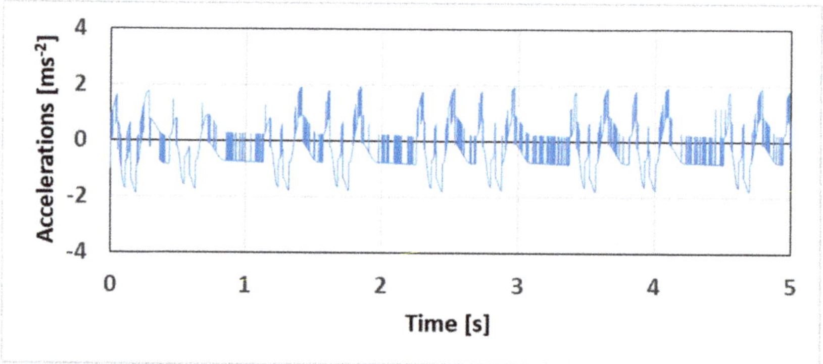

Figure 7. Accelerations of the flat wagon carrying structure in the center of gravity: bogie Y25.

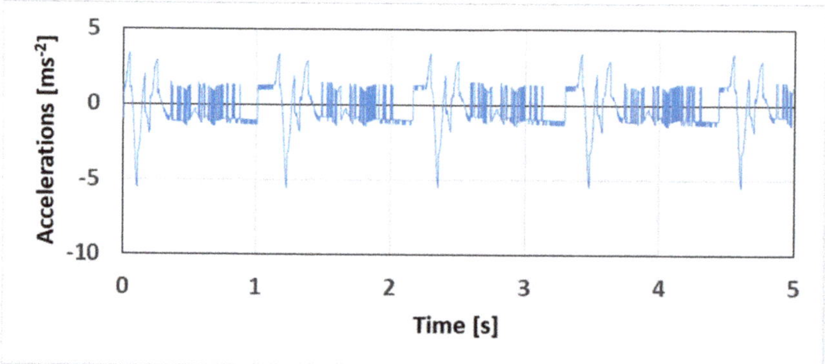

Figure 8. Accelerations of the flat wagon carrying structure in the bogie support areas: bogie 18–100.

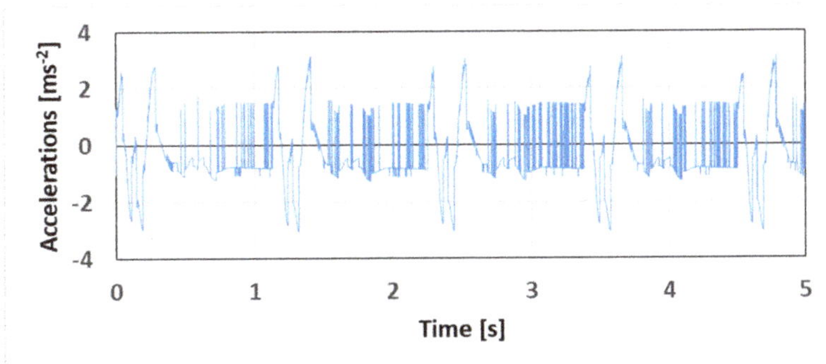

Figure 9. Accelerations of the flat wagon carrying structure in the bogie support areas: bogie Y25.

Mathematical model (1)–(6) was used for determination of other dynamic parameters of a flat wagon (Table 2). The calculation was made for a wagon speed of 80 kph.

Table 2. Dynamic parameters of an empty flat wagon in motion.

Parameter	Bogie Type		
	18–100	Y25	Parameter Improvement, %
Bogie acceleration, m/s^2	2.96	1.95	34.1
Bogie acceleration in areas of support on the bogie, m/s^2	5.53	3.16	42.9
Force in the spring suspension of a bogie, kN	38.5	21.53	44.2
Dynamic coefficient of the bogie	0.49	0.28	42.9

The results obtained made it possible to conclude that the dynamic parameters were within the allowable values and the motion of the wagon can be estimated as excellent [19,20]. A comparative study of the dynamic parameters of the flat wagon is given in Figure 10.

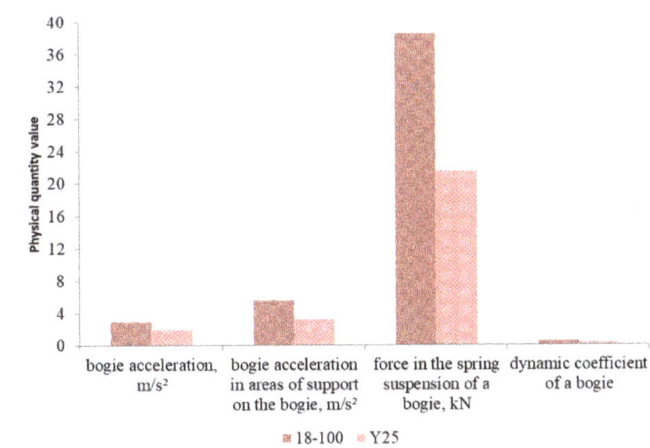

Figure 10. Comparative study of the dynamic parameters of the flat wagon with the 18–100 and Y25 bogies.

The application of the Y25 bogie for a flat wagon can decrease the acceleration of the carrying structure in comparison to that with the 18–100 bogie by 34%. The other dynamic parameters were also improved (Figures 11 and 12).

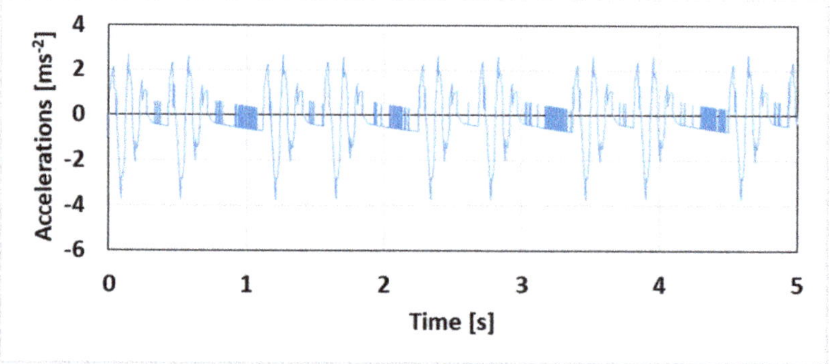

Figure 11. Accelerations of the carrying structure of the flat wagon in the center of gravity: bogie 18–100.

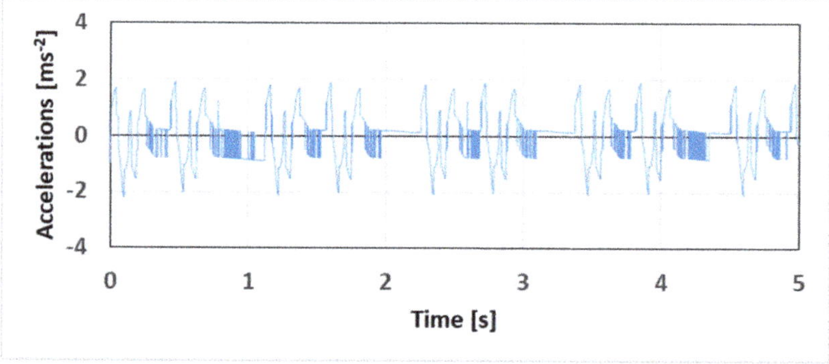

Figure 12. Accelerations of the carrying structure of the flat wagon in the center of gravity: bogie Y25.

The next research stage included determination of the dynamic characteristics of the flat wagon with the actual parameters. The accelerations on the carrying structure of the flat wagon in the center of gravity are given in Figures 11 and 12 and the accelerations in the areas of support on the bogies in Figures 13 and 14.

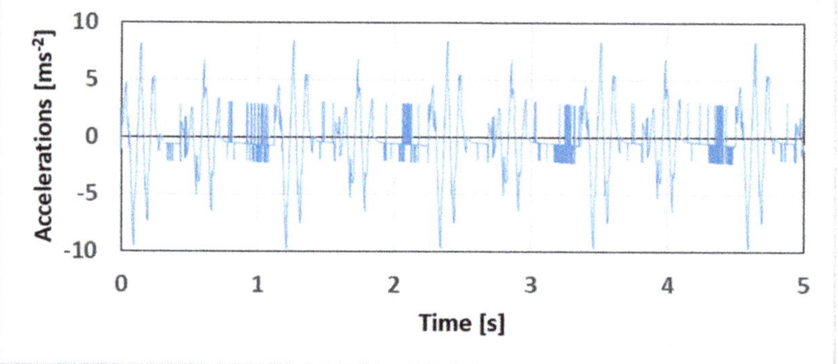

Figure 13. Accelerations of the flat wagon carrying structure in the bogie support areas: bogie 18–100.

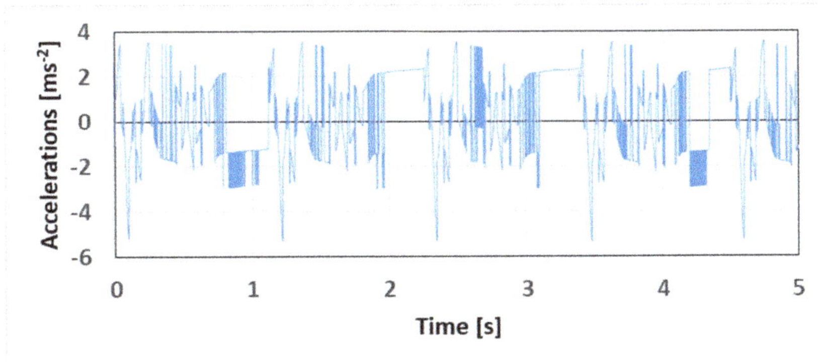

Figure 14. Accelerations of the flat wagon carrying structure in the bogie support areas: bogie Y25.

The mathematical model (1)–(6) was also used for determination of other dynamic parameters of the flat wagon (Table 3). The calculation was made for a wagon speed of 80 kph.

Table 3. Dynamic parameters of an empty flat wagon in motion.

Parameter	Bogie Type		
	18–100	Y25	Parameter Improvement, %
Bogie acceleration, m/s^2	3.72	2.21	40.6
Bogie acceleration in areas of support on the bogie, m/s^2	9.75	3.62	62.8
Force in the spring suspension of a bogie, kN	33.6	18.5	45.1
Dynamic coefficient of a bogie	0.61	0.34	44.3

The results obtained made it possible to conclude that the dynamic parameters were within the allowable values and the motion of the wagon can be estimated as excellent [19,20].

The comparative study of the dynamic parameters obtained for the flat wagon is given in Figure 15. It demonstrates improvements in percentage for certain dynamic parameters of the wagon with the Y25 bogie in comparison to those for a flat wagon with the 18–100 bogie.

The application of the Y25 bogie for the flat wagon with the actual parameters can decrease the acceleration of the carrying structure in comparison to that with the 18–100 bogie by 41%. The other dynamic parameters were also improved (see Figure 15).

The design service life of the flat wagon with the 18–100 and Y25 bogies was determined using the method presented in [21]:

$$T_n = \frac{(\sigma_{-1L}/[n])^m N_0}{B\, f_d \sigma_{ae}^m}, \tag{13}$$

where σ_{-1L}—average value of the endurance limit; n—allowable strength factor; m—fatigue curve exponent; N_0—test base; B—coefficient characterizing continuous work in seconds; f_d—efficient frequency of dynamic stresses; σ_{ae}—amplitude of equivalent dynamic stresses. The coefficient that characterizes the time of continuous operation of the wagon is determined by the formula:

$$B = \frac{365{\cdot}10^3 L_{avr}}{v_{avr}(1+0.34)}, \tag{14}$$

where L_{avr}—average daily wagon mileage, km (≈250 km); v_{avr}—average value of a wagon speed, m/s; 0.34—empty operating ratio.

The effective frequency of dynamic loadings is determined by

$$f_e = \frac{1.1}{2\pi}\sqrt{\frac{g}{y_s}}, \tag{15}$$

where y_s—static deflection of spring suspension, mm. The rest of the variables of Formula (13) are taken from the source [21].

The amplitude of equivalent dynamic stresses was determined by the formula:

$$\sigma_{ae} = \sigma_{wl}(k_{dv} + \psi_\sigma / K_\sigma), \tag{16}$$

where σ_{wl}—stresses from the static weight load; k_{dv}—coefficient of vertical dynamics; ψ_σ—sensitivity coefficient; K_σ—overall fatigue strength reduction coefficient.

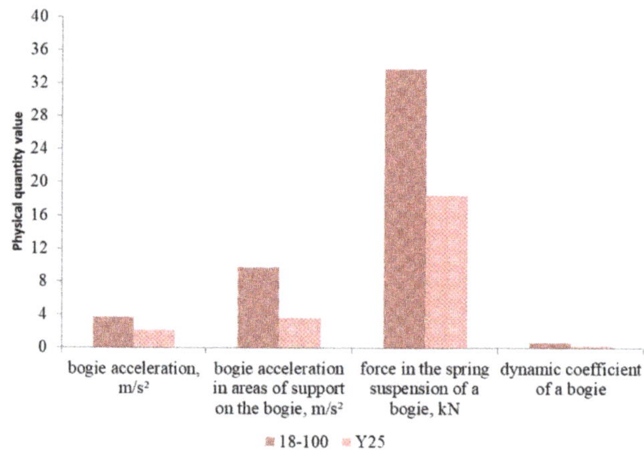

Figure 15. Comparative study of the dynamic parameters for a flat wagon with the 18–100 and Y25 bogies.

The determination of the amplitude of equivalent dynamic stresses included the side force coefficient equaled 1.1.

The following input parameters were taken for the calculation: σ_{1L} = 245 MPa; n = 2; m = 8; N_0 = 107; $B = 3.07 \times 10^6$ s; f_d = 2.7 Hz; ψ_σ / K_σ = 0.2.

To determine the stresses from the static weight load of the supporting structure of the flat wagon, a strength calculation was carried out using the finite element method in the SolidWorks simulation software package (CosmosWorks) [22–25]. Spatial isoparametric tetrahedrons were taken as the finite elements. Based on the previous research work of the authors, this element type provides sufficient convergence with technical experiments. Therefore, this type of tetrahedron was also selected in this study. The optimal number of elements was calculated with the graphic analytical method [26,27]. In this case, the mesh was created with respect to the curved surfaces of the supporting elements of the frame. In the areas of interfaces, as well as the interactions of structural elements with each other, the mesh was compacted using software options.

The number of elements in a mesh was 368,732, and number of nodes 14,938. The maximum element size in a mesh was 235.62 mm, the minimum size—47.12 mm, the maximum element side ratio—332; the percentage of elements with the side ratio less than three—24.6, and more than ten—31.5. 09G2S steel was taken as the material for the carrying structure, the mechanical parameters of which are given in Table 4.

The model was fixed in the areas of support on the bogies. The design diagram of the flat wagon is given in Figure 16. The study included the vertical load Pv, conditioned by the deadweight, on the carrying structure.

Table 4. Basic mechanical properties of 09G2S steel.

Property	Value
Elastic modulus, MPa	2.1×10^5
Ultimate strength, MPa	490
Yield strength, MPa	345
Shear modulus, MPa	7.9×10^4
Poisson's ratio Mass density, t/m^3	0.28
Mass density, t/m^3	7.8

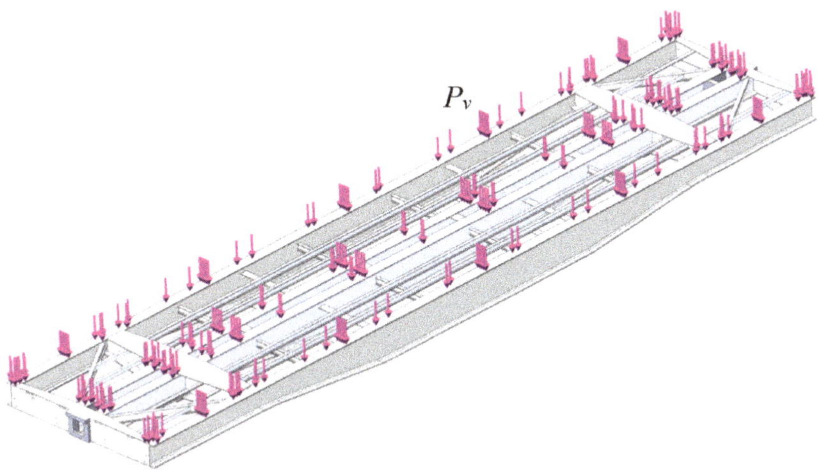

Figure 16. Design diagram of the wagon carrying structure.

The calculation demonstrated that the maximum equivalent stresses in the carrying structure of the flat wagon with the design parameters were concentrated in the contact area between the bolster beam and the center sill; they accounted for 88.95 MPa (see Figure 17). The carrying structure of the flat wagon with the actual parameters had the maximum equivalent stresses 79.95 MPa.

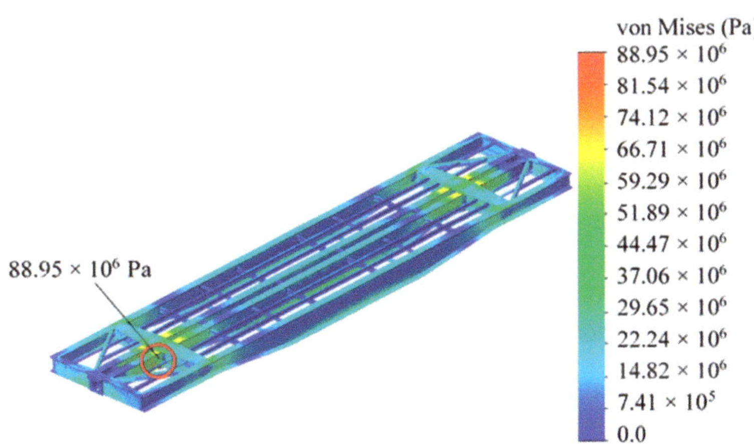

Figure 17. Stress state of the flat wagon carrying structure with the design parameters under the static weight load.

The calculation demonstrated that the design service life of the carrying structure of the flat wagon with the design parameters on the Y25 bogie twice long as the period calculated for the 18–100 bogie. At the same time, the maximum equivalent stresses in the supporting structure of the flat wagon with nominal parameters on bogies Y25 amounted to 88.95 MPa, and 170.3 MPa on bogies 18–100. Similar results were obtained for a flat wagon with the actual parameters. It should be noted that the service life obtained should be adjusted with consideration of additional studies into the longitudinal load on the carrying structure of the flat wagon and both field and bench experiments.

5. The Discussion of the Results Obtained

The authors made the mathematical modelling of dynamic loading for the 18–100 and Y25 bogies. The research was conducted for the 13–401 flat wagon with the design parameters and the actual parameters registered during field tests. It was found that application of the Y25 bogie can decrease the dynamic loading of the carrying structure of the flat wagon by more than 40% in comparison to that with the 18–100 bogie (Figure 16). The results of calculating the strength of the supporting structure of the flat wagon made it possible to conclude that the maximum equivalent stresses considering its nominal parameters on Y25 bogies, amounted to 88.95 MPa (Figure 17), which is half below the maximum equivalent stresses of the supporting structure on 18–100 bogies. The service life of the flat wagon can be prolonged twice. The calculation was made for an empty flat wagon in motion.

It should be noted that the authors studied the dynamic loading of the flat wagon in the vertical plane. They believe that further research in the field should determine the dynamic loading in the longitudinal and transverse directions. In addition, it is of primary importance to research the dynamic loading of the carrying structure of the flat wagon. It can be made with the strain gauge technique.

In addition, the support diagram of the Y25 bogie for flat wagons used on the broad-gauge lines should be improved. The authors will study these issues in their further research in the field.

6. Conclusions

The research deals with the mathematical modelling of the dynamic loading on the carrying structure of the flat wagon with the design parameters on the 18–100 and Y25 bogies. The approach was made in the plane coordinates for an empty wagon passing over a joint irregularity. The study found that application of the Y25 bogie for a flat wagon with the design parameters can decrease the dynamic load by 34% in comparison to that with the 18–100 bogie.

The authors made the mathematical modelling of the dynamic loading on the carrying structure of the flat wagon with the actual parameters on the 18–100 and Y25 bogies. The study included the actual parameters of the carrying elements of the flat wagon frame recorded during field tests. It was found that application of the Y25 bogie for a flat wagon with the actual parameters can decrease the dynamic loading by 40.6% in comparison to that with the 18–100 bogie.

The study found the design service life of the carrying structure of the flat wagon with the Y25 bogie. The calculation demonstrated that the design service life of the carrying structure of the flat wagon with the design parameters on the Y25 bogie is twice as long as the period obtained for the 18–100 bogie. Similar results were obtained for a flat wagon with the actual parameters.

This research can be of interest for those who are concerned about improvements in the dynamic characteristics and the strength (fatigue strength) of freight wagons, safe rail transportation, freight security, and development of new rail vehicles.

Author Contributions: Conceptualization, O.F. and A.L.; methodology, O.F.; software, V.P.; validation, A.L., V.P., and P.K.; formal analysis, A.L.; investigation, P.K.; resources, O.F.; data curation, V.P.; writing—original draft preparation, O.F.; writing—review and editing, A.L.; visualization, V.P.;

supervision, P.K.; project administration, A.L.; funding acquisition, A.L. All authors have read and agreed to the published version of the manuscript.

Funding: The authors gratefully acknowledge funding from the specific research on BUT FSI–S–20–6267.

Institutional Review Board Statement: Not applicable.

Informed Consent Statement: Not applicable.

Data Availability Statement: Not applicable.

Acknowledgments: The publication is part of the project "Development of conceptual measures for renovation of efficient operation for used freight cars". The project registration number is 2020.02/0122. The project is funded by the state budgetary institution National Research Foundation of Ukraine. The authors thank the Technologies, and Brno University of Technology for support.

Conflicts of Interest: The authors declare no conflict of interest.

References

1. Abood, K.H.A.; Khan, R.A. Investigation to improve hunting stability of railway carriage using semi–active longitudinal primary stiffness suspension. *J. Mech. Eng. Res.* **2020**, *2*, 97–105.
2. Buonsanti, M.; Leonardi, G. Dynamic modelling of freight wagon with modified bogies. *Eur. J. Sci. Res.* **2012**, *86*, 274–282.
3. Dižo, J.; Harušinec, J.; Blatnický, M. Structural analysis of a modified freight wagon bogie frame. In Proceedings of the MATEC Web of Conferences, České Budějovice, Czech Republic, 19 October 2017. [CrossRef]
4. Dižo, J.; Blatnický, M.; Pavlík, A. Process of modelling the freight wagon multibody system and analysing its dynamic properties by means of simulation computations. In Proceedings of the MATEC Web of Conferences, Strečno, Slovakia, 11–12 October 2018; Volume 235, pp. 1–8. [CrossRef]
5. Sun, Y.Q.; Cole, C. Vertical dynamic behavior of three–piece bogie suspensions with two types of friction wedge. *Multibody Syst. Dyn.* **2008**, *19*, 365–382. [CrossRef]
6. Hiensch, M.; Burgelman, N.; Hoeding, W.; Linders, M.; Steenbergen, M.; Zoeteman, A. Enhancing rail infra durability through freight bogie design. *Int. J. Veh. Mech. Mobil.* **2018**, *56*, 1532–1551. [CrossRef]
7. Fomin, O.; Lovska, A. Improvements in passenger car body for higher stability of train ferry. *Eng. Sci. Technol. Int. J.* **2020**, *23*, 1455–1465. [CrossRef]
8. Fomin, O.; Lovska, A. Establishing patterns in determining the dynamics and strength of a covered freight car, which exhausted its resource. *East. Eur. J. Enterp. Technol.* **2020**, *6*, 21–29. [CrossRef]
9. Fomin, O.; Lovskaya, A.; Plakhtiy, A.; Nerubatsky, V. The influence of implementation of circular pipes in load–bearing structures of bodies of freight cars on their physico–mechanical properties. *Sci. Bull. Nat. Mining Univ.* **2017**, *6*, 89–96.
10. Fomin, O.; Lovska, A.; Radkevych, V.; Horban, A.; Skliarenko, I.; Gurenkova, O. The dynamic loading analysis of containers placed on a flat wagon during shunting collisions. *ARPN J. Eng. Appl. Sci.* **2019**, *14*, 3747–3752.
11. Fomin, O.; Lovska, A.; Píštěk, V.; Kučera, P. Research of stability of containers in the combined trains during transportation by railroad ferry. *MM Sci. J.* **2020**, *1*, 3728–3733. [CrossRef]
12. Fomin, O.; Lovska, A.; Píštěk, V.; Kučera, P. Dynamic load computational modelling of containers placed on a flat wagon at railroad ferry transportation. In Proceedings of the Vibroengineering Procedia, Delhi, India, 28–30 November 2019; pp. 118–123. [CrossRef]
13. Lovska, A.; Fomin, O. A new fastener to ensure the reliability of a passenger coach car body on a railway ferry. *Acta Poly Tech.* **2020**, *60*, 478–485. [CrossRef]
14. Fomin, O.; Lovska, A. Determination of dynamic loading of bearing structures of freight wagons with actual dimensions. *East. Eur. J. Enterp. Technol* **2021**, *2/7(110)*, 6–15. [CrossRef]
15. Fomin, O.; Lovska, A.; Píštěk, V.; Kučera, P. Dynamic load effect on the transportation safety of tank containers as part of combined trains on railway ferries. In Proceedings of the Vibroengineering Procedia, Delhi, India, 28–30 November 2019; pp. 124–129. [CrossRef]
16. Aleksandrowicz, P. Verifying a truck collision applying the SDC method. In Proceedings of the 58th International Conference of Machine Design Departments, Praque, Czech Republic, 6–8 September 2017; pp. 14–19.
17. Kostek, R.; Aleksandrowicz, P. Effect of contact parameters on the pattern of vehicle collisions with a round pillar. In Proceedings of the 23rd International Conference Engineering Mechanics, Svratka, Czech Republic, 15–18 May 2017; pp. 490–493.
18. Aleksandrowicz, P. The impact of a vehicle braking system state on safe driving—Part two. In Proceedings of the 10th International Scientific Session Applied Mechanics, Bydgoszcz, Poland, 23 November 2018. [CrossRef]
19. DSTU 7598: 2014. Freight Cars. General Requirements for Calculations and Design of New and Modernized Cars of 1520 mm Gauge (Non–Self–Propelled). Kyiv, 2015. p. 162. Available online: http://online.budstandart.com/ua/catalog/doc-page.html?id_doc=73763 (accessed on 30 April 2021). (In Ukrainian).
20. GOST 33211–2014. Freight Cars. Requirements for Durability and Dynamic Qualities. Moscow, 2016. p. 54. Available online: https://docs.cntd.ru/document/1200121493 (accessed on 30 April 2021). (In Russian).

21. Ustich, P.A.; Karpych, V.A.; Ovechnikov, M.N. Reliability of Rail Non–Traction Rolling Stock. Moscow, 1999. p. 415. Available online: https://raillook.com/materialy/transport/jeleznodorojniy/vagony-i-vagonnoe-hozyaistvo/2004-nadezhnost-relsovogo-netyagovogo-podvizhnogo-sostava-p-a-ustich-v-a-karpyshev-m-n-ovechnikov/ (accessed on 30 April 2021). (In Russian)
22. Plakhtii, O.; Nerubatskyi, V.; Mashura, A.; Hordiienko, D. The analysis of mathematical models of charge–discharge characteristics in lithium–ion batteries. In Proceedings of the 2020 IEEE 40th International Conference on Electronics and Nanotechnology (ELNANO), Kyiv, Ukraine, 22–24 April 2020; pp. 635–640.
23. Plakhtii, O.; Nerubatskyi, V.; Sushko, D.; Hordiienko, D.; Khoruzhevskyi, H. Improving the harmonic composition of output voltage in multilevel inverters under an optimum mode of amplitude modulation. *East. Eur. J. Enterp. Technol.* **2020**, *2*, 17–24. [CrossRef]
24. Plakhtii, O.A.; Nerubatskyi, V.P.; Hordiienko, D.A.; Khoruzhevskyi, H.A. Calculation of static and dynamic losses in power IGBT–transistors by polynomial approximation of basic energy characteristics. *Sci. Bull. Nat. Mining Univ.* **2020**, *2*, 82–88. [CrossRef]
25. Otipka, V.; Zajac, R.; Prokop, A.; Řehák, K. Modal properties diagnostics of the high-pressure fuel injection pipes in off-road diesel engine. *J. Meas. Eng.* **2021**, *9*, 48–57. [CrossRef]
26. Vatulia, G.L.; Lobiak, O.V.; Deryzemlia, S.V.; Verevicheva, M.A.; Orel, Y.F. Rationalization of cross–sections of the composite reinforced concrete span structure of bridges with a monolithic reinforced concrete roadway slab. In Proceedings of the IOP Conference Series: Materials Science and Engineering, Sozopol, Bulgaria, 10–12 September 2019; pp. 1–9. [CrossRef]
27. Vatulia, G.; Komagorova, S.; Pavliuchenkov, M. Optimization of the truss beam. Verification of the calculation results. In Proceedings of the MATEC Web of Conferences, Kharkiv, Ukraine, November 14–16 2018; pp. 1–8. [CrossRef]

applied sciences

MDPI

Article

Evaluation of the Thermal Performance and Energy Efficiency of CRAC Equipment through Mathematical Modeling Using a New Index COP WEUED

Alexandre F. Santos [1,2], Pedro D. Gaspar [1,3] and Heraldo J. L. de Souza [2,*]

1 Department of Electromechanical Engineering, University of Beira Interior, 6201-001 Covilhã, Portugal; d1682@ubi.pt (A.F.S.); dinis@ubi.pt (P.D.G.)
2 FAPRO—Professional College, Curitiba 80230-040, Brazil
3 C-MAST—Centre for Mechanical and Aerospace Science and Technologies, 6201-001 Covilhã, Portugal
* Correspondence: heraldo@escolaprofissional.com.br; Tel.: +55-41-999748928

Abstract: As the world data traffic increasingly grows, the need for computer room air conditioning (CRAC)-type equipment grows proportionally. The air conditioning equipment is responsible for approximately 38% of the energy consumption of data centers. The energy efficiency of these pieces of equipment is compared according to the Energy Standard ASHRAE 90.1-2019, using the index Net Sensible Coefficient Of Performance (NetSCOP). This method benefits fixed-speed compressor equipment with a constant inlet temperature air-cooled condenser (35 °C). A new method, COP WEUED (COP–world energy usage effectiveness design), is proposed based on the IPLV (integrated part load value) methodology. The IPLV is an index focused on partial thermal loads and outdoor temperature data variation for air intake in the condenser. It is based on the average temperatures of the USA's 29 major cities. The new method is based on the 29 largest cities worldwide and with data-center-specific indoor temperature conditions. For the same inverter compressor, efficiencies of 4.03 and 4.92 kW/kW were obtained, using ASHRAE 90.1-2019 and the proposed method, respectively. This difference of almost 20% between methods is justified because, during less than 5% of the annual hours, the inlet air temperature in the condenser is close to the NetSCOP indication.

Keywords: ASHRAE 90.1; data center; IPLV; Net Sensible COP; AHRI 1361

Citation: Santos, A.F.; Gaspar, P.D.; de Souza, H.J.L. Evaluation of the Thermal Performance and Energy Efficiency of CRAC Equipment through Mathematical Modeling Using a New Index COP WEUED. *Appl. Sci.* **2021**, *11*, 5950. https://doi.org/10.3390/app11135950

Academic Editors: Hassane Naji, María Isabel Lamas Galdo and Rodriguez J.D.

Received: 21 May 2021
Accepted: 23 June 2021
Published: 26 June 2021

Publisher's Note: MDPI stays neutral with regard to jurisdictional claims in published maps and institutional affiliations.

Copyright: © 2021 by the authors. Licensee MDPI, Basel, Switzerland. This article is an open access article distributed under the terms and conditions of the Creative Commons Attribution (CC BY) license (https://creativecommons.org/licenses/by/4.0/).

1. Introduction

The Cisco Annual Internet Report forecasts the global adoption of the Internet. The proliferation of devices/connections and network performance, by the year 2023, will be [1]:

- 5.3 billion internet users (66% of the estimated population in 2023).
- 3.6 global devices and connections per capita.
- Average global speed of fixed broadband of 110 Mbps.
- In North America, 92% of the population will use the Internet.

In addition to the increasing number of users, there have been systems improvements, such as lower response time to search information, lower downtime (online for longer, without problems at critical moments), upgrade without interruption (in one click, management of active and unlimited resources), resilience and self-repair (makes the data pulverized automatically, and its update process is much simpler and optimized).

To sustain this growth in the global database and in user demand, the number of data centers and their energy consumption have been increasing. In 2018, it was estimated that data centers consumed 1% of all the global electricity generated. From 2010 to 2018, the number of computers skyrocketed (Figure 1) [2].

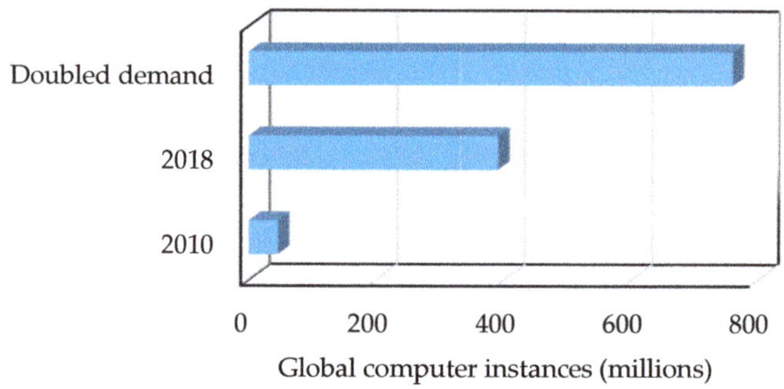

Figure 1. Growth of global data center instances [2].

Despite this growth, it is estimated that due to the increase in efficiency from 2010 to 2018, the total energy to serve the data centers grew by only 6%, and this is directly explained by the better efficiency of the data center equipment. The energy consumption of a data center is 10 to 100 times greater than that of a standard commercial building of the same dimensions. In Leadership in Energy and Environmental Design (LEED) buildings, the average energy consumption with an electrical load of 68% of the buildings was 10.8 W/m^2 [3]. According to the American Society of Heating, Refrigerating, and Air-Conditioning Engineers (ASHRAE), data centers are installations with an enormous demand for energy, highlighting the relevance of the theme. High-density data centers can reach 10,764 W/m^2, although on average, their consumption ranges from 430 to 861 W/m^2 [4]. Even though specific information technology (IT) equipment has evolved, within the data center, air conditioning systems are one of the main sinks of energy consumption. If energy consumption in data centers is considered of high relevance, the air conditioning topic becomes an indispensable item of discussion. On average, air conditioning systems are responsible for 38% of the energy consumption of data centers [5]. According to Santos et al. [6], the load distribution of a data center is distributed as shown in Figure 2.

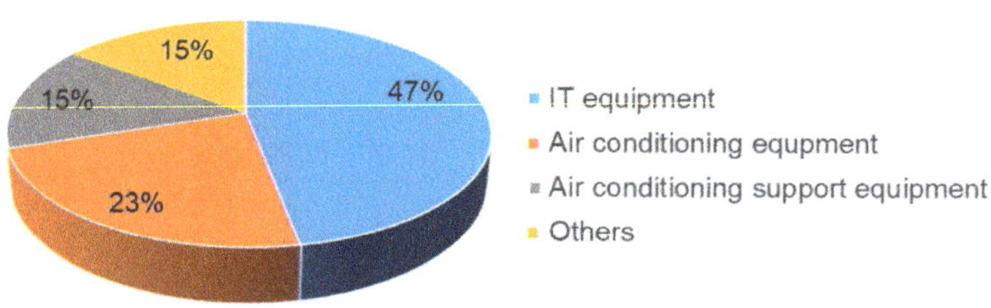

Figure 2. Distribution of electricity consumption in a typical DC with power usage effectiveness (PUE) = 2.1 [6].

Analyzing the current methodologies to measure the energy efficiency of computer room air conditioning (CRAC) equipment in data centers is an important action since it is the largest load apart from the IT equipment itself. Thus, suggesting a new methodology to measure the thermal performance and energy efficiency that considers the characteristics of the data center location is relevant in terms of sustainability.

2. Current Methodology (ASHRAE 90.1-2019 Standard)

A data center is different from ordinary commercial facilities, as it has a high sensible heat rate. Rack coolers are better if designed only for a sensible rate (without wet coils), and the coils can even be above the dew temperature [7].

Based on this high sensible heat rate, wisely, the ASHRAE 90.1-2019 Standard uses the Net Sensible Coefficient Of Performance (NetSCOP) index as a basis for the user to compare the efficiency of the air conditioning machine with a minimum of efficiency specified in the standard. This condition is important because, in some medium data centers, standard equipment (self-contained air conditioning) can be used, and split-type air conditioning system may be used in some smaller data centers [8]. Different from common equipment, the CRAC is designed specifically for data centers. The project type of downflow air supply is used often, and CRAC has the appearance of a closet, designed for a high sensible heat rate, providing more reliability. Figure 3 shows a sectional view of CRAC equipment installed in a data center, and Figure 4 shows a plan view of CRAC installation within a downflow safe room.

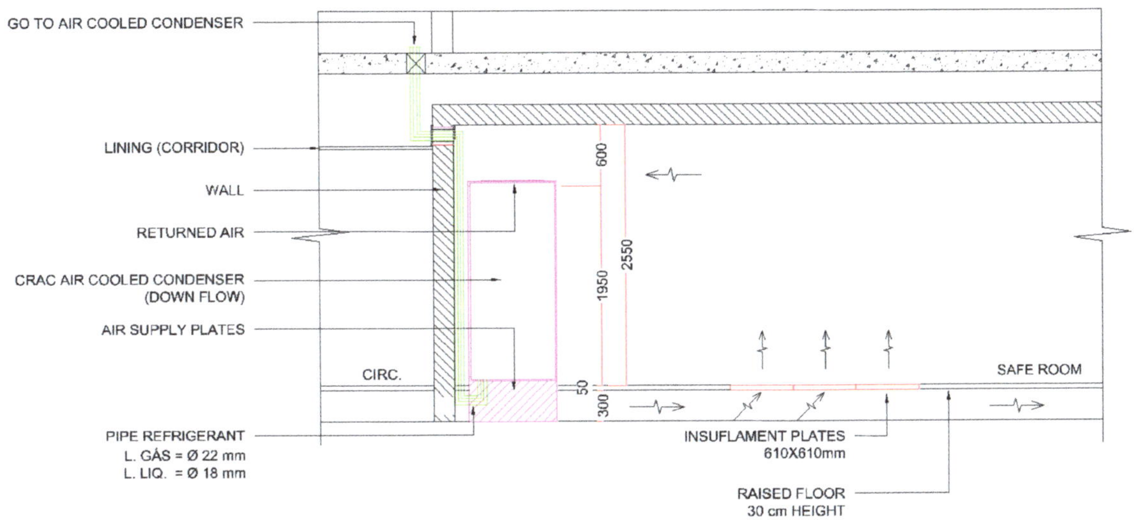

Figure 3. Sectional view of CRAC equipment installed in a data center.

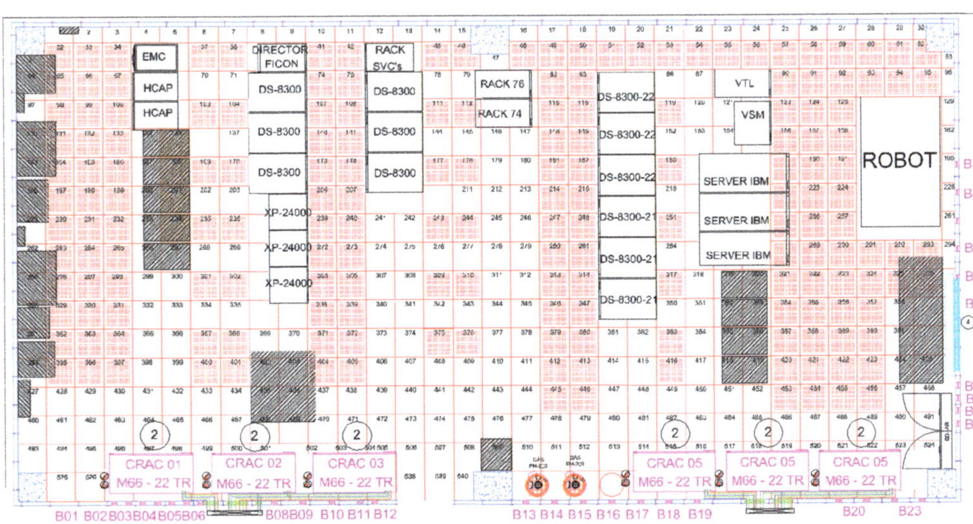

Figure 4. Plan view of CRAC installation within a downflow safe room.

Table 1 shows the minimum efficiency requirements and the Net Sensible COP in ASHRAE 90.1-2019 Standard indicated for floor-mounted air conditioners and condensing units serving computer rooms [8] considering the rating conditions for dry-bulb (DBT) and dew-point (DPT) temperatures.

Table 1. Minimum efficiency requirements and the Net Sensible COP in ASHRAE 90.1-2019 Standard indicated for floor-mounted air conditioners and condensing units serving computer rooms [8].

Equipment Type	Standard Model	Net Sensible Cooling Capacity COP	Minimum Net Sensible COP (kW/kW)	Rating Conditions Return Air (DBT/DPT)	Test Procedure
Air-cooled	Downflow	<23 kW	2.70	29 °C/11 °C (Class 2)	AHRI 1361
		≥23 and <86 kW	2.58		
		≥86 kW	2.36		
	Upflow-duct	<23 kW	2.67		
		≥23 and <86 kW	2.55		
		≥86 kW	2.33		
	Upflow-nonduct	<19 kW	2.16	24 °C/11 °C (Class 1)	
		≥19 and <70 kW	2.04		
		70 kW	1.89		
	Horizontal	<19 kW	2.65	35 °C/11 °C (Class 3)	
		≥19 and <70 kW	2.55		
		70 kW	2.47		

The values for the most used equipment, which is the downflow, are based on a fixed characteristic of return temperature and dew point based on Class 2. According to TC 9.9, classes are divided according to the types of equipment/needs of a data center, as shown in Table 2 [9].

Table 2. Classes for certain characteristics of data center enclosures [9].

2008 Classes	2011 Classes	Applications	Information Technology Equipment	Environmental Control
1	**A1**	Data center	Enterprise servers, storage products	Tightly controlled
2	**A2**		Volume servers, storage products, personal computers, workstations, laptop, printers	Some control
	A3			Some control, use of free cooling techniques when allowable
	A4			Some control, near full-time usage of free cooling techniques
3	B	Office, home, transportable environment, etc.	Personal computers, workstations, laptops, printers	Minimal control
4	C	Point of sale, industrial, factory, etc.	Point of sale equipment, ruggedized controllers or computers, PDAs	No control

A1—A data center environment with strict control of the psychrometric parameters: dry-bulb temperature (DBT), dew-point temperature (DPT), relative humidity (RH), and in a mission-critical operation. Generally developed for large companies with a large number of racks.

A2—Generally a technological production environment or an office or a laboratory with some control over environmental parameters. They are locations that shelter small racks; they can be personal servers or workstations.

The values of return dry-bulb temperature (DBT) and dew-point temperature (DPT) of 29 and 11 °C, respectively (see Table 3), are recommended for Class 2. These values are not recommended for Class 1, but they could be allowable.

It is also important to note that a return temperature does not mean the intake of air in the rack, which will certainly be at a lower temperature.

Table 3. ASHRAE 2015 Thermal Guidelines classes [10].

Equipment Environmental Specifications for Air Cooling							
Product Operations						Product Power Off	
Class	**DBT (°C)**	**Humidity Range**	**MDP (°C)**	**Max. Elevation(m)**	**Max. Δ*T*/hour (°C/h)**	**DBT (°C)**	**RH (%)**
Recommend (suitable for all four classes)							
A1–A4	18 to 27	−9 °C (DPT) to 15 °C (DPT) and 60% RH					
Allowable							
A1	15 to 32	−12 °C (DPT) and 8% RH to 17 °C (DPT) and 80% RH	17	3050	5/20	5 to 45	8 to 80
A2	10 to 35	−12 °C (DPT) and 8% RH to 21 °C(DPT) and 80% RH	21	3050	5/20	5 to 45	8 to 80
A3	5 to 40	−12 °C (DPT) and 8% RH to 24 °C (DPT) and 85% RH	24	3050	5/20	5 to 45	8 to 80
A4	5 to 45	−12 °C (DPT) 8% RH to 24 °C (DPT) and 90% RH	24	3050	5/20	5 to 45	8 to 80
B	5 to 35	8% RH to 28 °C(DPT) and 80% RH	28	3050	N.A.	5 to 45	8 to 80
C	5 to 40	8% RH to 28 °C(DPT) and 80% RH	28	3050	N.A.	5 to 45	8 to 80

The parameters and methodologies to arrive at the values of ASHRAE 90.1-2019 are specified in AHRI 1361-2017. The air supply parameters of Table 4 of the standard of the Air-Conditioning, Heating, and Refrigeration Institute (AHRI) corroborate with ASHRAE 90.1-2019.

Table 4. Indoor return air temperature standard rating conditions [11].

Mounting Locations	Standard Model	Cooling (Return Air DBT/DPT) (°C)	Humidification (Return Air DBT/DPT) (°C)
Ceiling mounted unit	Ceiling mounted unit ducted	24.0/11.0	24.0/5.6
	Ceiling mounted unit nonducted		
Floor mounted unit	Upflow unit nonducted	24.0/11.0	
	Upflow unit ducted	29.5/11.0	
	Downflow unit	29.5/11.0	
	Horizontal flow unit	35.0/11.0	

In contrast, AHRI 1361-2017 also offers the air intake temperatures in the condenser, as shown in Table 5. It must be noted that:

1. All ratings are at standard atmospheric pressure.
2. For the NetSCOP calculation, add allowance for cooling tower fan(s) and heat rejection loop, water pump power input in kW to the unit total input in kW = 5% of the unit net sensible capacity.
3. For the NetSCOP calculation, add allowance for dry cooler fan(s) and heat rejection loop glycol pump power input in kW to the unit total input in kW = 7.5% of the unit net sensible capacity.
4. For the NetSCOP calculation, add allowance for chilled water pump power input in kW to unit total input in kW (See Equation (1)).

Table 5. Heat rejection/cooling fluid standard rating conditions [11].

System Type	Fluid Condition	Test Condition
Air-cooled units	Entering outdoor ambient DBT (°C)	35.0
Water-cooled units (typically connected to a common glycol loop)	Entering water temperature (EWT) (°C)	28.5
	Leaving water temperature (LWT) (°C)	35.0
	Water Flow rate (L/s)	N/A
Glycol-cooled units (typically connected to a common glycol loop)	Entering glycol temperature (°C)	40.0
	Leaving glycol temperature (°C)	46.0
	Glycol Flow rate (L/s)	N/A
	Glycol solutions concentration	40% Propylene glycol by volume
Chiller-water units (typically connected to a common chilled water loop)	Entering water temperature (EWT) (°C)	10.0
	Leaving water temperature (LWT) (°C)	16.5

It is appropriate that in a place of thermal load and air supply conditions constant at 8760 h, the priority is the efficiency in Net Sensible COP. However, although there is no variation internally, there is an external factor in the equipment: the temperature of the air inlet in the condenser constantly changes, implying a positive or negative change in the performance of the air conditioning equipment. This temperature variation has a relevant impact on the performance of the inverter or digital compressor equipment type.

The integrated part load value (IPLV) is a performance characteristic developed and used in AHRI's methodology. This methodology considers variable air intake temperature in the condenser and variable thermal load based on the average temperature of the 29 major cities in the United States of America (U.S.A.).

IPLV is a methodology in which COP is measured in partial loads. IPLV is a parameter to consider even in chillers for data centers. Its parameters are described in AHRI 550/590-2015 [12]. These parameters are in accordance with Equation (1) and described in Table 6.

$$\text{IPLV (or NPLV)} = 0.01 \cdot A + 0.42 \cdot B + 0.45 \cdot C + 0.12 \cdot D \quad (1)$$

where,

A: COP at 100% capacity, (kW/kW).
B: COP at 75% capacity, (kW/kW).
C: COP at 50% capacity, (kW/kW).
D: COP at 25% capacity, (kW/kW).

Table 6. Partial load conditions for calculating IPLV/NPLV.

Evaporator (all types)	Unit	Values
100% capacity leaving water temperature (LWT)	(°C)	6.7
Volumetric flow	(m^3/h·ton)	5.45
Fouling Factor	($m^{2\cdot}$ °C/W)	0.000018
Water condenser		
100% capacity entering water temperature (EWT)	(°C)	29.4
75% capacity EWT	(°C)	23.9
50% capacity EWT	(°C)	18.3
25% capacity EWT	(°C)	18.3
0% capacity EWT	(°C)	18.3
Volumetric flow	(m^3/h·ton)	6.81
Fouling Factor	($m^{2\cdot}$ °C/W)	0.000044
Air condenser		
100% capacity entering water temperature (EWT)	(°C)	35.0
75% capacity EWT	(°C)	26.7
50% capacity EWT	(°C)	18.3
25% capacity EWT	(°C)	12.8

From a macro point of view, there is an index that analyzes the conditions of free cooling, evaporative system, and variable COP in data centers: the Energy Usage Effectiveness Design (EUED), with the following characteristics (see Figure 5) [13]:

- When the outside air temperature is below 20 °C and the enthalpy is below 42.7979 kJ/kg, only free cooling will be used.
- When the temperature is between 15 and 24 °C and the enthalpy is from 42.7979 to 55.8233 kJ/kg, the evaporative system will be used.
- When the temperature is above 20 °C and the enthalpy is over 55.8233 kJ/kg, the normal system will be used under the following conditions:
  1. Air intake temperature between 24.1 and 27 °C, called COP1.
  2. Air intake temperature between 27.1 and 30 °C, called COP2.
  3. Air intake temperature between 30.1 and 33 °C, called COP3.
  4. Air intake temperature above 33.1 °C in any condition, called COP4.
  5. If a geothermal temperature is available, it will be used to determine the COP, with a differential of 4 °C of the geothermal temperature.

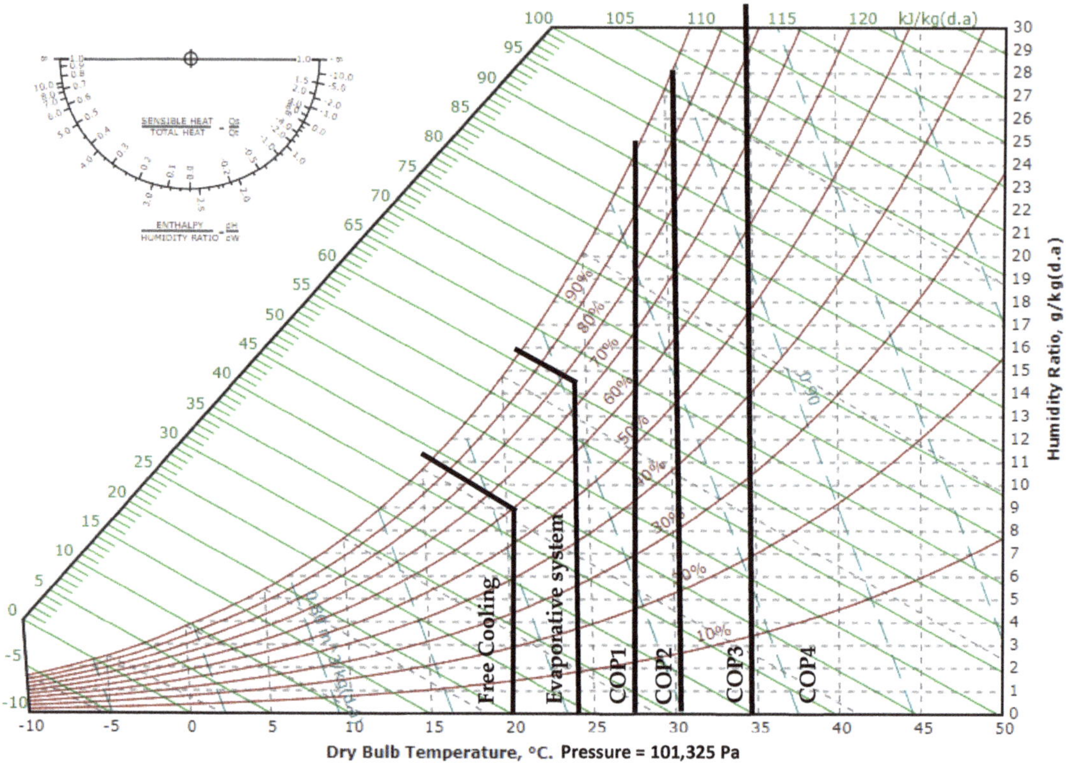

Figure 5. Psychrometric chart with variation of case studies.

3. New Methodology

The new methodology proposed in this paper considers the superposition of the three methodologies: Net Sensible COP, IPLV, and EUED, to create a specific index for CRAC equipment with an air-cooled condenser with downflow air supply named COP WEUED (COP World Energy Usage Effectiveness Design). This index emphasizes the CRAC equipment with the compressor on (i.e., active refrigeration cycle), knowing that indexes such as EUED or statistical analysis for predicting location-specific data center PUE and its improvement potential, already use free cooling and evaporative cooling for their methodologies. The air intake characteristics in the condenser must consider the use of the refrigeration cycle in the data center air conditioning system. Among the characteristics of the new index are [3,13]:

(1) Fixed air return temperature conditions equivalent to ASHRAE 90.1-2019, that is, 29 °C with a dew point temperature of 11 °C, thus considering a standard evaporation temperature of 12 °C.

(2) Air intake temperatures in the condenser calculated for four values (levels) shown in Table 7. Each of the values of COP1, COP2, and COP3 are the average values of the EUED methodology, whereas COP4 is the value used by AHRI 1361 for air intake in the condenser.

(3) Considering the principle in the IPLV that is based on the 29 largest cities in the U.S.A., in this case, the 29 largest cities in the world are used. Table 8 lists the 29 most populous cities according to the 2018 United Nations report [14].

Table 7. Air inlet temperatures in the condenser.

Level	Calculation Temperature (°C)
COP1	25.5 (COP1 average from 24.1 to 27 °C)
COP2	28.5 (COP2 average from 27.1 to 30 °C)
COP3	31.5 (COP3 average from 30.1 to 33 °C)
COP4	35 (All temperatures above 33.1 °C)

Table 8. World cities and their populations [14].

No.	City	Country	Population [×1000]
1	Tokyo	Japan	37,468
2	Delhi	India	28,514
3	Shanghai	China	25,582
4	São Paulo	Brazil	21,650
5	Mexico City	Mexico	21,581
6	Al-Qahirah-Cairo	Egypt	20,076
7	Mumbai	India	19,980
8	Beijing	China	19,618
9	Dhaka	Bangladesh	19,578
10	Osaka	Japan	19,281
11	New York	USA	18,810
12	Karachi	Pakistan	15,400
13	Buenos Aires	Argentina	14,967
14	Chongqing	China	14,838
15	Istambul	Turkey	14,751
16	Calcutta	India	14,681
17	Manila	Philippines	13,482
18	Lagos	Nigeria	13,462
19	Rio de Janeiro	Brazil	13,293
20	Tianjin	China	13,215
21	Kinshasa	Democratic Republic Congo	13,171
22	Guangzhou	China	12,638
23	Los Angeles	USA	12,458
24	Moscow	Russia	12,410
25	Shenzhen	China	11,908
26	Lahore	Pakistan	11,738
27	Bangalore	India	11,440
28	Paris	France	10,901
29	Bogota	Colombia	10,574

The ASHRAE Weather Data Viewer [15] was used to determine the average temperature condition of each of the 29 largest cities in the world. Table 9 shows how many hours in each of these cities in the world are for COP1, COP2, COP3, and COP4. It is important to remember that the hours of free cooling and evaporative cooling will not be part of the refrigeration equipment index with the compression cycle.

Table 9. COPs of cities worldwide [15].

City	Country	Annual Hours COP1 (24 to 26.9 °C)	Annual Hours COP2 (27 to 29.9 °C)	Annual Hours COP3 (30 to 32.9 °C)	Annual Hours COP4 (≥33 °C)
Tokyo	Japan	910	674	284	57
Delhi	India	879	1481	1323	1873
Shanghai	China	1190	817	340	125
São Paulo	Brazil	1.215	454	147	13
Mexico City	Mexico	2.610	1669	1168	1020
Al-Qahirah-Cairo	Egypt	1345	947	785	660
Mumbai	India	1915	3412	1844	468
Beijing	China	1051	526	262	86
* Dhaka	Bangladesh	1561	2.685	1.487	794
Osaka	Japan	959	849	416	130
New York	USA	545	260	95	18
Karachi	Pakistan	1201	2476	1848	1053
Buenos Aires	Argentina	1083	389	85	9
Chongqing	China	743	743	437	309
Istambul	Turkey	830	443	126	15
Kolkata (Calcutta)	India	1561	2685	1487	794
Manila	Philippines	1959	4436	2024	262
* Lagos	Nigeria	1345	1762	1708	2686
Rio de Janeiro	Brazil	2807	1533	494	91
Tianjin	China	955	668	321	94
* Kinshasa	Dem. Rep. Congo	2966	1746	997	216
Guangzhou	China	1737	1685	822	341
Los Angeles	USA	198	63	18	4
Moscow	Russia	205	85	23	6
Shenzhen	China	1708	2049	911	138
Lahore	Pakistan	985	1320	1103	1525
Bangalore	India	2164	1156	520	156
Paris	France	257	113	42	12
Bogota	Colombia	2	0.2	0.04	0
TOTAL		36,886	37,126	21,117	12,955
Average		1271.92	1280.20	728.17	446.72
Total Hours		3727.02			

Note: * Specifically, the ASHRAE Weather Data Viewer [15] has temperatures in 26 of the 29 cities. For the cities of Dhaka, Lagos, and Kinshasa values from nearby cities, Calcutta, Niamey, Brazzaville, respectively, were used.

As shown in Table 9, using the weighted average of the 29 largest cities in the world, the use of compression refrigeration in data centers is essential in 3727.2 h per year of the 8760 h available. That is, it is feasible to use free cooling and evaporative cooling for the other 5032.8 h per year, as shown in Figure 6.

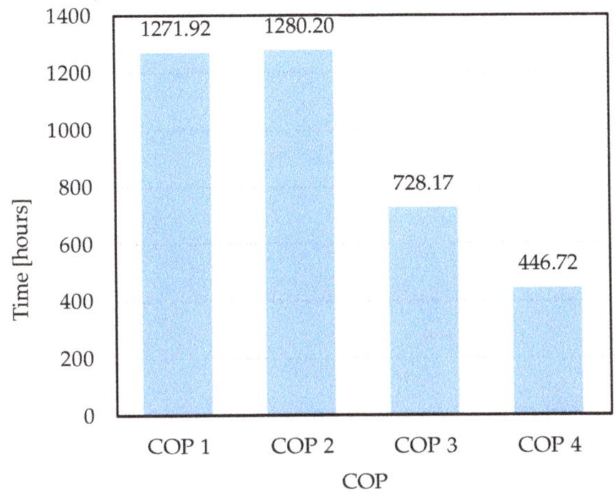

Figure 6. COPs vs. annual hours of refrigeration system usage.

Using part of the IPLV formula, EUED, and AHRI 13621 concepts, Equation (2) was determined using percentages of COPs shown in Figure 7 [16].

$$\text{COP WEUED} = ((0.34 \cdot \text{COP1}) + (0.34 \cdot \text{COP2}) + (0.20 \cdot \text{COP3}) + (0.12 \cdot \text{COP4})) \cdot \text{SLR} \quad (2)$$

where SLR is the sensible heat rate.

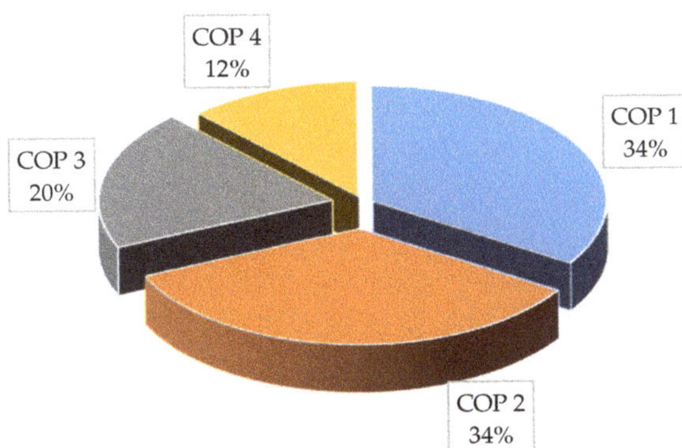

Figure 7. COPs percentages.

4. Analysis and Discussion

To demonstrate experimentally the difference between a system with both conditions, a calculation using the Bitzer Inverter Scroll compressor software is firstly developed for the AHRI 1361 condition and then with the COP WEUED condition. The considerations exposed in Table 10 were used.

AHRI conditions and thermal load:

- Sensible cooling capacity = 50 kW;
- Total cooling capacity = 55 kW;
- Inlet air condenser temperature = 35 °C (AHRI 1361 conditions) [16];
- Approach between bubble temperature and condenser air inlet = 10 °C;
- Condensing temperature = 45 °C;
- Evaporating temperature = 10 °C;
- Suction gas superheat = 10 °C (EN 12900-2013 conditions);
- Liquid subcooling (in condenser) = 0 °C (EN 12900-2013 conditions) [17].

COP Sensible WEUED conditions:

The conditions are the same as the AHRI conditions except for the air inlet temperature in the condenser. The comparison of results is shown in Table 10 for the compressor Bitzer GSD60137VA4.

Table 10. AHRI vs. COP WEUED result comparison.

	AHRI	COP1	COP2	COP3	COP4
Compressor freq. (Hz)	73	64	67	69	73
Cooling capacity (kW)	55	55	55	55	55
Evaporator capacity (kW)	55	55	55	55	55
Condenser capacity (kW)	67.5	64.1	65.6	66	67.5
COP/EER (kW/kW)	4.43	5.98	5.46	4.97	4.43
Min. cooling capacity (kW)	26.4 (35 Hz)	329.6 (35 Hz)	28.6 (35 Hz)	27.7 (35 Hz)	26.4 (35 Hz)
Max. cooling capacity (kW)	56.4 (75 Hz)	63.5 (75 Hz)	61.3 (75 Hz)	59 (75 Hz)	56.4 (75 Hz)
Mass flow (kg/h)	1240	1115	1159	1186	1240
Discharge gas temp. w/o cooling (°C)	76.7	64.2	68	71.9	76.7
Result COP (kW/kW)	4.03	4.92			

The COP WEUED is determined using Equation (2):

$$\text{COP WEUED} = ((0.34 \times 5.98) + (0.34 \times 5.46) + (0.2 \times 4.97) + (0.12 \times 4.43)) \times 0.91 = 4.92 \text{ kW/kW}$$

As can be analyzed from results, while the Net Sensible COP with AHRI 1361 conditions is 4.43 kW/kW (but with sensible heat equal to 4.03 kW/kW), the proposed COP WEUED index value is 5.41 kW/kW (but with sensible heat equal to 4.92 kW/kW). A considerable difference of 19% is determined, due to:

(a) With the AHRI 1361 method for CRAC equipment, it is impossible to show the difference between fixed compressors and inverter for data centers in COP evaluations.
(b) It has been proven that even at fixed thermal loads, there is an advantage of an air conditioning system with a variable flow of refrigerant fluid (inverter system).
(c) Just as in AHRI, there are the IPLV and NPLV that use the same formula. The main difference between them is that IPLV is based on AHRI characteristics and NPLV is based on local characteristics. COP WEUED can also be used based on local characteristics. For example, using the same methodology for the city of São Paulo, Equation (2) provides:

$$\text{COP NEUED} = ((0.66 \times 5.98) + (0.25 \times 5.46) + (0.08 \times 4.97) + (0.01 \times 4.43)) \times 0.91 = 5.23 \text{ kW/kW}$$

That is, in the case of São Paulo, the difference would be 23% to COP NEUED vs. COP WEUED. While the current NetSCOP method value was 4.03 kW/kW, COP WEUED was 4.92 kW/kW, and with specific data from the city of São Paulo, COP NEUED was 5.23 kW/kW, all simulated with the same inverter compressor.

5. Conclusions

Despite the evolution of data centers in reducing energy consumption, the index used to measure and compare energy efficiency between CRAC equipment does not yet use inverter technology resources (variable refrigerant flow) in its methodology. The IPLV for equipment already was developed for comfort air conditioning, but the Net Sensible COP methodology favors fixed-capacity equipment. The energy efficiency index needs to keep up with new technologies; according to Wen et al. [18], the compressor frequency variation is one of the greatest technologies for reducing energy consumption in CRAC equipment. Both data center equipment and air conditioning systems are evolving. Another technology is microchannels coils with microfluids that can reduce heat dissipation from IT equipment with both air and water cooling [19]. However, microchannel coils can also be used in air conditioning equipment and improve energy savings [20]. The COP WEUED index measures more accurately the benefits of these new technologies.

Just as the IPLV is an important milestone for air conditioning equipment, a COP WEUED index was created based on the 29 major cities in the world, which could be an important tool to compare CRAC equipment, gathering the best of the AHRI 1361, which prioritizes sensible heat, with the calculation of the EUED method and also with the IPLV formula.

This methodology can be useful for further studies, as it can serve as a basis for manufacturing CRAC equipment with more connection to the real temperatures of the outside air, even recalculating the condenser fans, since the specific mass of the air is also related to temperatures.

In addition to these advantages, the proposed method favors high-performance air conditioning equipment in the range in which they will be truly used, since technologies such as free cooling and evaporative cooling are already realities in data centers.

Author Contributions: Conceptualization, A.F.S. and P.D.G.; methodology, A.F.S.; validation, A.F.S. and H.J.L.d.S.; formal analysis, A.F.S. and P.D.G.; investigation, A.F.S.; resources, H.J.L.d.S.; data curation, A.F.S. and P.D.G.; writing—original draft preparation, A.F.S. and H.J.L.d.S.; writing—review and editing, P.D.G.; visualization, H.J.L.d.S.; supervision, P.D.G.; project administration, A.F.S. All authors have read and agreed to the published version of the manuscript.

Funding: This research received no external funding.

Institutional Review Board Statement: Not applicable.

Informed Consent Statement: Not applicable.

Data Availability Statement: Not applicable.

Acknowledgments: Authors acknowledge Fundação para a Ciência e a Tecnologia (FCT—MCTES) for its financial support via the project UIDB/00151/2020 (C-MAST).

Conflicts of Interest: The authors declare no conflict of interest.

Abbreviations

CRAC	Computer room air conditioner.
ASHRAE	American Society of Heating, Refrigeration, and Air Conditioning Engineers.
COP	Coefficient of performance.
SLR	Sensible heat rate.
EUED	Energy Usage Effectiveness Design.
WEUEDN	World Energy Usage Effectiveness Design Nonstandard.
IPLV	Integrated part load value.
LEED	Leadership in Energy and Environmental Design.
IT	Information technologies.
TC	Technical committee.
AHRI	Air-conditioning, Heating, and Refrigeration Institute.
NPLV	Non-standard part load value.
PUE	Power usage effectiveness

References

1. CISCO. By The Numbers. Projecting the Future of Digital. 2020. Available online: https://www.cisco.com/c/en/us/solutions/executive-perspectives/annual-internet-report/infographic-c82-741491.html (accessed on 12 October 2020).
2. Masanet, E.; Shehabi, A.; Lei, N.; Smith, S.; Kooney, J. Recalibrating global data center energy-use estimates. *Science* **2020**, *367*, 984–986. [CrossRef] [PubMed]
3. Lei, N.; Masanet, E. Statistical analysis for predicting location-specific data center PUE and its improvement potential. *Energy* **2020**, *201*, 117556. [CrossRef]
4. Amiri, A.; Ottelin, J.; Sorvari, J. Are LEED-Certified Buildings Energy-Efficient. *Sustainability* **2019**, *11*, 1672. [CrossRef]

5. Ni, J.; Bai, X. A review of air conditioning energy performance in data centers. *Renew. Sustain. Energy Rev.* **2017**, *67*, 625–640. [CrossRef]
6. Santos, A.F.; Gaspar, P.D.; Souza, H.J.L.d. New Data Center Performance Index: Perfect Design Data Center—PDD. *Climate* **2020**, *8*, 110. [CrossRef]
7. NREL. *High-Performance Computing Data Center*; National Renewable Energy Laboratory (NREL), U.S. Department of Energy: Denver, CO, USA, 2014.
8. ASHRAE. *Standard 90.1-2019 (SI Edition)—Energy Standard for Buildings Except Low-Rise Residential Buildings (ANSI Approved; IES Co-sponsored)*; American Society of Heating, Refrigerating and Air-Conditioning Engineers (ASHRAE): Atlanta, GA, USA, 2019.
9. ASHRAE. *ASHRAE TC 9.9—Data Center Networking Equipment—Issues and Best Practices*; American Society of Heating, Refrigerating and Air-Conditioning Engineers (ASHRAE): Atlanta, GA, USA, 2016.
10. ASHRAE. *ASHRAE—Thermal Guidelines Classes*; American Society of Heating, Refrigerating and Air-Conditioning Engineers (ASHRAE): Atlanta, GA, USA, 2015.
11. AHRI. *AHRI 1360. Performance Rating of Computer and Data Processing Room Air Conditioners*; Air-Conditioning, Heating, and Refrigeration Institute (AHRI): Arlington, VA, USA, 2017.
12. AHRI. *AHRI Standard 550/590 (I-P). Performance Rating of Water-Chilling and Heat Pump Water-Heating Packages Using the Vapor Compression Cycle*; Air-Conditioning, Heating, and Refrigeration Institute (AHRI): Arlington, VA, USA, 2015.
13. Santos, A.F.; Gaspar, P.D.; Souza, H.J.L. Evaluation of the Heat and Energy Performance of a Datacenter Using a New Efficiency Index: Energy Usage Effectiveness Design—EUED. *Braz. Arch. Biol. Technol.* **2019**, *62*, e19190021. [CrossRef]
14. United Nations Department of Economic and Social Affairs, Population Division. *The World's Cities in 2018*; ST/ESA/SER.A/417 Data Booklet; UN: New York, NY, USA, 2018.
15. ASHRAE. *ASHRAE Weather Data Viewer, Version 5.0*; American Society of Heating, Refrigerating and Air-Conditioning Engineers (ASHRAE): Atlanta, GA, USA, 2016.
16. AHRI. *AHRI Standard 1361. Standard for Performance Rating of Computer and Data Processing Room Air Conditioners*; Air-Conditioning, Heating, and Refrigeration Institute (AHRI): Arlington, VA, USA, 2017.
17. ES. *DIN EN 12900—Refrigerant Compressors—Rating Conditions, Tolerances and Presentation of Manufacturer's Performance Data*; European Standard (ES), Deutsches Institut fur Normung E.V. (DIN): Berlin, Germany, 2013.
18. Wen, J.; Wenlun, C.; Bei, C.; Yuyao, H. An inverter testing system for CRAC. In Proceedings of the 2014 International Conference on Intelligent Green Building and Smart Grid (IGBSG), Taipei, Taiwan, 23–25 April 2014; pp. 1–5. [CrossRef]
19. Cairone, F.; Gagliano, S.; Bucolo, M. Experimental study on the slug flow in a serpentine microchannel. *Exp. Therm. Fluid Sci.* **2016**, *76*, 34–44. [CrossRef]
20. Park, C.Y.; Hrnjak, P. Experimental and numerical study on microchannel and round-tube condensers in a R410A residential air-conditioning system. *Int. J. Refrig.* **2008**, *31*, 822–831. [CrossRef]

applied sciences

MDPI

Article

CFD Investigations of Cyclone Separators with Different Cone Heights and Shapes

Satyanand Pandey [1], Indrashis Saha [2], Om Prakash [1], Tathagata Mukherjee [1], Jawed Iqbal [3], Amit Kumar Roy [4], Marek Wasilewski [5,*] and Lakhbir Singh Brar [1]

1 Mechanical Engineering Department, Birla Institute of Technology, Mesra, Ranchi 835215, India; pandeysatyanand007@gmail.com (S.P.); omprakash@bitmesra.ac.in (O.P.); tatha95@gmail.com (T.M.); brarls@bitmesra.ac.in (L.S.B.)
2 Chemical Engineering Department, University of Calcutta, Kolkata 700073, India; indrashissaha98@gmail.com
3 Civil and Environmental Engineering Department, Birla Institute of Technology, Mesra, Ranchi 835215, India; jawed@bitmesra.ac.in
4 Mechanical Engineering Department, Sikkim Manipal University, Majitar, Rangpo 737136, India; amittuntun@gmail.com
5 Faculty of Production Engineering and Logistics, Opole University of Technology, 76 Proszkowska St., 45-758 Opole, Poland
* Correspondence: m.wasilewski@po.opole.pl

Abstract: Due to the great achievements in the field of optimization of the design of cyclone separators, non-standard solutions are sought to increase their performance. Therefore, in this study, we consider the impact of different cone and cylinder height variants on the performance of cyclone separators. Additionally, we propose non-standard shapes for these sections. Three different heights: $H/D = 0.5$, 1.0, and 1.5, with D (the main cyclone body diameter), are analyzed. Since the cone is one of the most important geometrical entities, three different shapes viz. a straight (conventional) profile, a concave profile as well as a convex profile are also taken into account. Cyclone performance is rated at three different inlet velocities viz. $U_{in} = 10$ m/s, 15 m/s, and 20 m/s. Hence, a total of 27 simulations have been performed using the Reynolds stress model. It becomes apparent from the present study that the pressure loss is lowest in the convex variant, whereas the separation efficiency is better in the conventional design. Furthermore, an increase in the length of the cylindrical section reduces pressure drop with a mild decrease in the collection efficiency in all variants.

Keywords: cyclone separator; separation efficiency; pressure drop; convex and concave cyclones; Reynolds stress model

Citation: Pandey, S.; Saha, I.; Prakash, O.; Mukherjee, T.; Iqbal, J.; Roy, A.K.; Wasilewski, M.; Brar, L.S. CFD Investigations of Cyclone Separators with Different Cone Heights and Shapes. *Appl. Sci.* **2022**, *12*, 4904. https://doi.org/10.3390/app12104904

Academic Editors: Hassane Naji, María Isabel Lamas Galdo and Rodriguez J.D.

Received: 20 April 2022
Accepted: 9 May 2022
Published: 12 May 2022

Publisher's Note: MDPI stays neutral with regard to jurisdictional claims in published maps and institutional affiliations.

Copyright: © 2022 by the authors. Licensee MDPI, Basel, Switzerland. This article is an open access article distributed under the terms and conditions of the Creative Commons Attribution (CC BY) license (https://creativecommons.org/licenses/by/4.0/).

1. Introduction

Cyclone separators are mechanical separators that are used to remove the particulate matter contained in the gas stream. Due to tangential injection and the curved wall, the dust-laden gas experiences a swirling motion that generates a centrifugal force field readily felt by suspended solid particles. This causes the solid particles to move in a radially outward direction towards the walls, thereby aiding the separation process. Hence, the particle concentration near the walls is higher for larger particles, and these particles are transported down to the collection bin attached at the cyclone bottom under the influence of axial velocity [1]. Cyclones are used in almost every industry where the solid particles are to be separated before releasing the fluid into the open atmosphere. Cyclones are used either as pre-separators to reduce the dust loading on the bag filters or as a final separation unit. Other roles include the classification of particles according to the size to recover the catalyst and many more. These devices are popular due to several advantages such as significantly shorter particle residence time, simple construction, ability to operate continuously even in corrosive environments, lower maintenance cost, and no moving parts (to name a few) [2,3]. Cyclones find application in several industries such as steel plants, cement industries, flour mills, power plants, etc. [4].

Since the development of cyclone separators, several studies have been undertaken to optimize the construction of these separators. The vortex finder diameter significantly affects the performance of cyclone separators—with an increase in diameter of the vortex finder, the pressure drop largely reduces, followed by a net reduction in the collection efficiency [5,6] and vice-versa. The insertion length of the outlet tube mildly influences the cyclone performance [7,8]. The eccentricity provided to the centre of the outlet tube largely affects the cyclone performance [9]. If performed without special care, the cyclone performance may deteriorate; hence, this aspect requires performing an optimization before implementation [10]. The vortex finder with cross-shaped blades decreases the pressure drop by more than 16% with a slight improvement of 0.64% in the separation efficiency [11]. Installing the de-swirler closer to the vortex finder inlet counts more on cyclone performance [12]. The convergent-divergent vortex finder helped to improve the pressure drop by as much as 6% followed by a reduction in the cut size from 1.6 μm to 1.4 μm [13]. The main cyclone body diameter (*D*) significantly influences the performance—an increase in the cyclone main body diameter increases the pressure drop, whereas the collection efficiency of larger particles is improved [14,15]. The increase in the length of cyclones plays a vital role; in fact, this is the only parameter that at the same time improves the performance indicators [16]. Apart from the inlet dimensions [17], the bend angle of the inlet duct [6,18] and the location of the secondary inlet [19] likewise affect its performance. The increasing value of the inlet angle was reported to increase the cut-off size and reduce the pressure drop coefficient [20]. The pressure drop in cyclone was reported to be the least at an inlet angle of 450 and the collection efficiency was found to improve with further increase in this angle [21]. A contracted inlet duct length largely affects cyclone performance [22]—increasing the length of the duct enhanced the tangential velocity. The outlet tubes with inverted cone shapes were noticed to yield a higher pressure drop and collection efficiency [23]. The notched outlet tubes were reported to increase the overall cyclone performance [24]. The location of the multi-inlet duct was found as an important parameter [25]. Compared to the widely used tangential inlet, a scroll-type inlet design attaching several subsidiary cyclones to the exterior of the cyclone body increased collection efficiency and significantly lowered the pressure drop [26]. The circular rod inserted at the centre of cyclones was reported to enhance cyclone performance [27]. The dipleg shape also affected the pressure drop and collection efficiency [28]—the pressure drop was highest observed in cylindrical dipleg, whereas the lowest pressure drop occurs in the cyclone without a dipleg. A cyclone separator with fin diameter ratios of 2/5 or porosity equal to 0.2 was recommended for lower pressure drops [29]. Qiang et al. [30] considered optimizing the performance of two-stage cyclones in series using response surface methodology and reported enhancement in the overall performance. A dynamic cyclonic classifier was employed to classify particles according to their sizes by Galletti et al. [31], wherein a rotating impeller was employed to improve the efficiency of the cyclone. Another classifier was proposed by Sun et al. [32], in which three characteristic regions were used to capture solid particles based on the particle size. The novel slotted vortex finder was reported to enhance the cyclone performance by Fu et al. [33]. Mofarrah et al. [34] designed a cyclone for indoor use to separate the pollens that may lead to severe allergies in humans. An et al. [35] made use of cyclone separators for sampling black carbon.

The cone tip diameter was found to have a mild effect on the cyclone performance [17,36]. Xiang et al. [37] experimented on a lab-scale cyclone model with a diameter of 0.031 m to determine the pressure drop and collection efficiency. They reported a mild pressure drop as well as collection efficiency with a decrease in the cone tip diameter. However, Stern et al. [38] (mentioned in [37]) stated that the cone was merely a carrier of separated solid particles, and that it does not have any influence on the cyclone performance. In a different study by Shastri and Brar [39] and Shastri et al. [40], the cone was revealed to be an important entity in cyclones. With an increase in the cone length (thus the overall height of the cyclone is the same), the collection efficiency was found to increase, and vice versa—the cyclone model without the cone was reported to have the lowest collection efficiency. Brar et al. [16] reported that an increase in the length of the conical section significantly increases the collection efficiency than an increase in the length of the cylindrical section.

Considering the studies performed so far on the conical section, it becomes apparent that the cone is one of the most important parts of the cyclone. However, it has also been observed that the impact of different shapes of the conical walls on cyclone performance has not been taken into account so far. Therefore, in this study, we consider studying the impact of the curved wall profile of the conical section rather than the straighter one (as in the conventional design) for analysis. Here, two different shapes viz. the cone with concave wall and the convex wall profiles are accounted for. The definition of convexity and concavity of the walls is in context to the observer exterior to the cyclone. A part of the convex profile and concave profile has been earlier analyzed by [41] in hydrocyclone, and no such study has been undertaken in the context of the gas cyclones. Hence, the present study undertakes different shapes of the cone section. Apart from the different shapes of the cone section, the length of the cylindrical section is also included in this study to evaluate its impact on the different shapes of cone sections.

In short, we consider three different lengths of the cylindrical section while three different shapes of the conical section viz. a conventional (straight) profile, concave profile, and convex profile for analysis. The length of the cylinder is varied such that the total height of the cyclone remains the same. The cyclone performance has been analyzed through numerical simulations. The pressure drops, separation efficiencies and cut-off particle sizes of the new variants have been compared to the conventional design to evaluate the relative differences.

2. Materials and Methods

For analysis and optimization of cyclone separators, the CFD approach is widely preferred over the experiment. This is due to the expensive nature of the experimental setup, and optimization would demand fabricating several cyclones. In such cases, CFD has proven as the best alternative to the expensive experimental methods, and this technique has gained tremendous popularity worldwide. Therefore, the present study makes use of CFD to predict the performance indicators of cyclones. For this, we make use of the commercial CFD code Fluent—a popular finite volume solver.

2.1. Continuous Phase Equations

Given that the flow is incompressible and isothermal, the continuity, and the Reynolds-Average Navier-Stokes (RANS) equation can be adequately expressed as [42]:

$$\frac{\partial u_i}{\partial x_i} = 0 \tag{1}$$

$$\rho\frac{\partial u_i}{\partial t} + \rho u_j\frac{\partial u_i}{\partial x_j} = -\frac{\partial P}{\partial x_i} + \frac{\partial u_i}{\partial x_j}\left[2\mu S_{ij} - \rho\overline{u_i' u_j'}\right] \tag{2}$$

where u_i—the mean velocity; x_i.the coordinate system, P—mean pressure, ρ—density of the gas, and μ is the dynamic viscosity of the continuous phase. Equation (2) determines the Reynolds stress tensor, $\tau_{ij} = -\rho\overline{u_i' u_j'}$

$$\text{The strain rate tensor, } S_{ij} = \frac{1}{2}\left[\frac{\partial u_i}{\partial x_j} + \frac{\partial u_j}{\partial x_i}\right] \tag{3}$$

Reynolds Stress Model (RSM)

The way to solve the Reynolds stress tensor τ_{ij} is most important for numerical studies. In the RSM closure model, all six Reynolds stress components are solved along with the dissipation rate. Ignoring the production conditions related to the buoyancy and rotation of the system, the transport equations for RSM can be defined as [42]:

$$\frac{\partial \tau_{ij}}{\partial t} + u_k \frac{\partial \tau_{ij}}{\partial x_k} = -\tau_{ik} \frac{\partial u_j}{\partial x_k} - \tau_{jk} \frac{\partial u_i}{\partial x_k} + \epsilon_{ij} - \phi_{ij} + \frac{\partial}{\partial x_k} \left[\nu \frac{\partial \tau_{ij}}{\partial x_k} + C_{ijk} \right] \tag{4}$$

The right part of the Equation (4) represents the response: derivative of local time and convection. In turn, the left side represents the stress generation, the dispersion tensor, and the pressure-strain correlation tensor, respectively. The first and second components in parentheses represent molecular diffusion and turbulent diffusion tensors.

$$\text{The dissipation tensor, } \epsilon_{ij} = -2\nu \overline{\frac{\partial u_i'}{\partial x_k} \frac{\partial u_j'}{\partial x_k}} \tag{5}$$

$$\text{The pressure-strain correlation tensor, } \phi_{ij} = \overline{\frac{p'}{\rho} \left[\frac{\partial u_i'}{\partial x_j} + \frac{\partial u_j'}{\partial x_i} \right]} \tag{6}$$

$$\text{The turbulent diffusive transport tensor,} \quad \rho C_{ijk} = \rho \overline{u_i' u_j' u_k'} + \overline{p' u_i'} \delta_{jk} + \overline{p' u_j'} \delta_{ik} \tag{7}$$

The term δ, in Equation (7), is termed the Kronecker delta. Ignoring the pressure fluctuations, the tensor C_{ijk} is symmetrical in all three indexes, which makes it invariant in rotation. To close Equation (4), the tensors must be defined ϵ_{ij}, ϕ_{ij}, and C_{ijk} from Equations (5)–(7) [42,43].

2.2. Solid-Phase Transport

To define the trajectory of the solid phase, the particle force equilibrium equation is configured within the Lagrange frame of reference. The force balance equation can be written as follows:

$$\frac{du_{pi}}{dt} = F_D(u_i - u_{pi}) + \frac{(\rho_p - \rho)}{\rho_p} g_i + F_i \tag{8}$$

where: the subscript i denotes the direction of the flow parameter, u_{pi}—the particle velocity; F_i—additional force (e.g., Brownian force, Saffman lift, etc.) expressed as force per unit mass, the term $F_D(u_i - u_{pi})$. —drag force per particle mass unit; ρ_p—the particle density; g_i—the grav. acceleration.

For spherical particles, the drag force (Stokes) can be defined as follows:

$$F_D = \frac{18\mu}{\rho_p d_p^2} \frac{C_D Re}{24} \tag{9}$$

where: d_p—the diameter of the particle; Re—the Reynolds number; μ—the molecular viscosity of the fluid phase.

$$Re = \frac{\rho d_p |u_{pi} - u_i|}{\mu} \tag{10}$$

In Equation (8), the second term on the right-hand side viz. $C_D * Re/24$ approaches unity for Stokes flow. The first term bears a unit of reciprocal of time and is a representation of the particle response time, given as:

$$\tau_p = \frac{\rho_p d_p^2}{18\mu} \tag{11}$$

The drag coefficient for smooth particles is given as (a_1, a_2, and a_3—constants) [44]:

$$C_D = a_1 + \frac{a_2}{Re_p} + \frac{a_3}{Re_p^2} \tag{12}$$

A single particle velocity is determined by integrating Equation (8) in discrete time steps, and its position in the non-stationary flow is determined from the equation:

$$\frac{d_{xi}}{d_t} = u_{pi} \tag{13}$$

Then the Equation (8) can be expressed in the following form:

$$\frac{du_{pi}}{dt} = \frac{1}{\tau_p}(u_i - u_{pi}) + a_i \tag{14}$$

where: the term a_i contains accelerations due to all force terms; τ_p—the particle relaxation time.

Fluent includes several options for tracing (or discretizing) the trajectory of solid particles. In most of the research on flow in cyclone separators carried out so far, higher-order schemes, especially trapezoidal ones, have been recommended [45].

In trapezoidal diagrams, the particle's progress in each successive discrete time step is determined. In Equation (12), the right part is averaged as:

$$\frac{u_{pi}^{n+1} - u_{pi}^{n}}{\Delta t} = \frac{1}{\tau_p}\left(u_i^* - u_{pi}^*\right) + a_i{}^n \tag{15}$$

where: u_{pi}^*, u_i^*—mean values of variables u_i, u_{pi}—whose values can be expressed as appropriate: $u_{pi}^* = \left(u_{pi}^n + u_{pi}^{n+1}\right)/2$, $u_i^* = \left(u_i^n + u_i^{n+1}\right)/2$, and $u_i^{n+1} = u_i^n + \Delta t u_{pi}^n \nabla u_i^n$.

The particle velocity at the new location $n + 1$ is expressed as:

$$u_{pi}^{n+1} = \frac{u_{pi}^n\left(1 - \frac{\Delta t}{2\tau_p}\right) + \left(u_i^n + \frac{1}{2}\Delta t u_{pi}^n \nabla u_i^n\right)\frac{\Delta t}{\tau_p} + \Delta t a}{1 + \frac{\Delta t}{2\tau_p}} \tag{16}$$

Since the solid particles used here are in the range of 0.4–8.0 microns and present in the dilute phase, it was necessary to take into account the turbulent dispersion effect caused by instantaneous velocity (defined as the sum of average velocity and variable velocity). For this reason, a stochastic tracking model (random walk) has been used to account for the effect of instantaneous velocity fluctuations on the trajectories of particles. The dispersion of turbulent particles is defined by integrating the trajectory equations for each solid particle in the particle path.

The discrete random walk (DRW) model—the stochastic model—has been used to account for the effect of the instantaneous velocity of the carrier fluid on the solid particles. The turbulence is assumed to be consistent with the Gaussian probability distribution in variable velocity using the correlation as:

$$u_i' = \zeta\sqrt{\overline{u_i' u_j'}} \tag{17}$$

with ζ as a normally distributed random number, and the remaining term on the right-hand side represents the local root mean square (RMS) of velocity fluctuations. With RSM, the characteristic lifetime of an eddy is taken as:

$$\tau_e = -T_L\, ln(r) \tag{18}$$

where r is a uniform random number ($0 < r < 1$), and the fluid Lagrangian integral time scale T_L is defined as:

$$T_L \approx 0.3k/\varepsilon \tag{19}$$

2.3. Details of the Analyzed Geometric Variants and the Adopted Assumptions of CFD Study

The performance of all the cyclone geometries considered in the present work is analysed using numerical simulations. The governing (differential) equations, which are in conservation form, are first integrated over each control volume. Next, these equations are converted into a set of linear algebraic equations, which are further solved iteratively with user-defined solver tolerance (residuals) and under-relaxation factors. Since there are no moving bodies or boundaries in the present case, the Eulerian grids are generated. However, since the secondary (particulate) phase is tracked throughout the flow field, this demands implementing the Lagrangian approach. Hence, collectively, the approach is termed the Euler-Lagrangian approach [40].

Since cyclonic flows constitute high streamlined curvatures with rapid changes in strain, and steep pressure gradients—which makes it highly anisotropic—a one equation model (viz. Spallart-Allmaras), and two-equation models (such as k-ε and k-ω closures) are not suitable here due to isotropic treatment to the turbulence. Hence, a more advanced (a seven equation) closure model to RANS viz. Reynolds stress model (RSM) is preferred, which solves transport equations for the Reynolds stresses together with an equation for the dissipation rate. RSM is capable to account for the anisotropy associated with the swirling flows [9,13,36,40].

2.3.1. Details of the Cyclone Models

The Stairmand high-efficiency cyclone model [46] is used as a baseline model to evaluate the performance of the proposed cyclone design. The cyclone Stairmand is a cylinder-on-cone structure to which the inlet of the two-phase mixture (solid-fluid) is tangentially attached at the top. The geometry of this separator includes the following sections: two-phase mixture inlet (defined by height a and width b); vortex finder diameter (D_e) and length (L_v); the diameter of the lower conical section (B_c); the height of the cylindrical section (H), the height of the conical section (H_c), the dipleg height (L_c), and cyclone diameter (D), as shown in Figure 1a; the dimensions are given in Table 1.

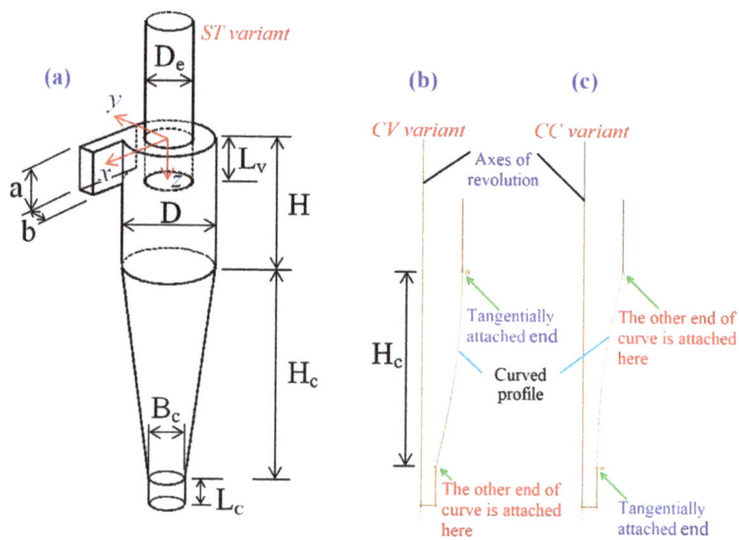

Figure 1. Geometrical details of the Stairmand high-efficiency cyclone model (D = 0.205 m) with (**a**) straight walls of the conical section (referred to as the ST), (**b**) convex profile (referred to as the CV variant), and (**c**) concave profile (referred to as the CC variant).

Table 1. Dimensions of the cyclone models.

Cyclone Variants	Cyclones Models	Symbol	Dimensionless Ratio (Dimension)/D
		D_e	0.5
		a	0.5
		b	0.2
		L_v	0.5
		L_c	0.5
		B_c	0.375
ST	ST1	H	0.5
	ST2 *		1.5
	ST3		2.5
CV	CV1		0.5
	CV2		1.5
	CV3		2.5
CC	CC1		0.5
	CC2		1.5
	CC3		2.5
ST	ST1	H_c	3.5
	ST2		2.5
	ST3		1.5
CV	CV1		3.5
	CV2		2.5
	CV3		1.5
CC	CC1		3.5
	CC2		2.5
	CC3		1.5

* Standard (Stairmand high efficiency) cyclone model with $D = 0.205$ m.

Apart from the straight walls of the conical section (referred to as the ST (straight) variant), we also consider curved conical walls. For this, the two proposed curved geometries include a convex profile (referred to as the CV variant, cf. Figure 1b), and a concave profile (referred to as the CC variant, cf. Figure 1c)—the profile description is based on the observations external to the cyclone. The convex profile, as shown in Figure 1b, consists of a parabolic curve attached tangentially to the vertical line of the cylinder at the top, while the other end is attached to a short downcomer tube. On the other hand, in the concave profile (cf. Figure 1c), one end is attached tangentially to the downcomer tube, whereas the other end is attached to the cylindrical section at the top. After completion of (2-D) geometry on the y plane, the cyclone model was generated by revolving the profile about the (z-) axis of revolution. As a final step, a rectangular duct was attached to all models near the cyclone roof.

In addition to the above, we further explore the impact of different heights of the cylindrical section on the cyclone performance—the height of the cylinder is changed while maintaining a constant total cyclone height ($H + H_c$). The reason for not permitting any change to the total height is to optimize the design for a compact size. Such an approach was considered earlier by authors [15,39,40]. The details of all the cyclone geometries are given in Table 1, and the final geometries of all the models are represented in Figure 2.

To summarize, in the present study, we categorize the cyclone variants based on the profile of the conical wall viz. a straight profile (ST variant), convex profile (CV variant) and concave profile (CC variant). The different heights of the cylindrical section in each variant are then further categorized as the cyclone models that contain a numeric character after the variant description (cf. Figure 2).

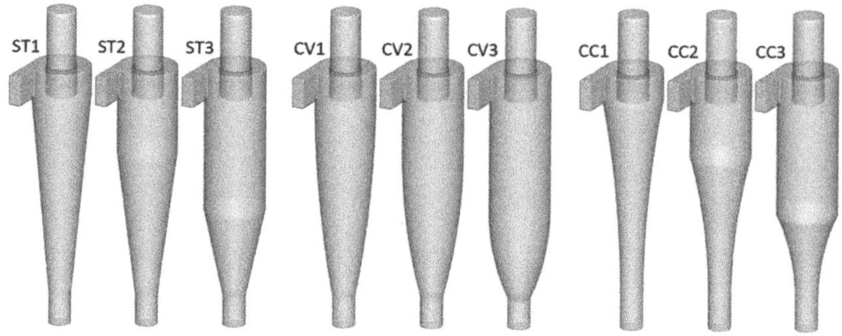

Figure 2. Structural representation of all the cyclone models of ST variant, CV variant and CC variant (from left to right), respectively.

2.3.2. Mesh Generation and Solver Settings

Since there are no moving bodies or boundaries in the present case, the Eulerian grids are generated using ICEM CFD; in this, the block-structured mesh was created that conforms to the curved boundaries. The generated hexahedral mesh is non-orthogonal, and the cell faces are well aligned to the direction of the fluid flow—this helps to reduce numerical diffusion. Three levels of the grid are considered viz. 214K, 386K and 568K, referred to as level 1, level 2, and level 3 mesh, respectively, and evaluated for grid dependence. Details of the surface mesh (of level 2) for the cyclone models ST2, CV2, and CC2 are illustrated in Figure 3. For all the cases, and all grid resolutions, the maximum aspect ratio was 32, and the mean skewness was 0.88 (near the central brick in the core region of the cyclone). At different grid resolutions, the velocity profiles were found to be nearly the same with no remarkable difference. The maximum difference in pressure drop was found to be less than 3.8%, and for cut-off particle size, the maximum difference was 3%.

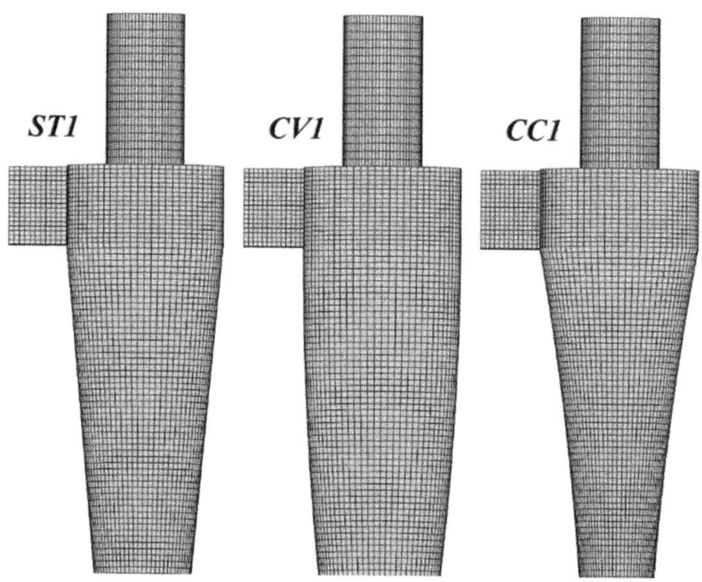

Figure 3. Sample mesh regions for all cyclone models ST1, CV1, and CC1. Three levels of the grid are considered viz. 214K, 386K and 568K (from left to right), respectively.

In the present study—assuming a low volume fraction of solid particles loading (less than 10% by volume)—we consider one-way coupling. This has been considered in numerous studies (e.g., [8–10,12,17,20]) without any loss in generality. The transient simulations were initialized with reference to the inlet velocity and zero gauge pressure at the inlet.

For pressure and velocity coupling, we use the SIMPLEC algorithm. For pressure discretization, we use the PRESTO! interpolation scheme. The momentum equations are discretized using the QUICK scheme. For turbulent kinetic energy and dissipation rate, the second-order upwind scheme is used. For a stable solution, the first-order upwind scheme is applied to Reynolds stresses. The solver tolerance is set to 1×10^{-5} for all the conservation equations. A time step size of 0.0001 s is used for the considered variants. To account for the near-wall effects, we use the standard wall function. Such numerical settings were also used by [16,47–50].

For the solid particles, 1,000,000 integral steps are used to keep track of the particle trajectories. A length scale of 0.005 m has been opted. To consider the effect of turbulence on solid particles in the flow field, the discrete random walk model has been applied. The characteristic lifetime of an eddy is accounted for using the random eddy lifetime model with a trapezoidal discretization scheme [16,17,51].

2.3.3. Materials and Boundary Conditions

For the continuous phase, air with a density of 1200 kg/m^3 and viscosity equal to 2×10^{-5} kg/m^{-s} has been used. The velocity inlet boundary condition ($\overline{U} = -U_{in}\hat{n}$.) [52,53] with turbulent intensity ($I = u'/U_{in}$; u' as the root mean square of the velocity) set to 5% is used at the inlet and a uniform velocity profile is applied. At the cyclone outlet, the pressure outlet boundary condition (with zero-gauge pressure) is applied. All other walls of the cyclone separator were assigned a *no-slip* boundary condition ($\boldsymbol{u} \cdot \hat{n} = 0.$) [52,53]. Three different inlet velocities viz. U_{in} =10, 15, and 20 m/s are considered in the present study to analyse the performance of all cyclone models.

The solid phase was talcum powder ρ_p = 2700 kg/m^3). For solid particles, the uniformly sized particles are considered with eight different sizes ranging from 0.4 to 8.0 μm—these particles are introduced from the inlet in the direction normal to the inlet face. The solids touching the bottom plate are considered to be collected by the cyclone; hence, the *trap* boundary condition is applied there. The *escape* boundary condition is applied at the outlet plane, and the solid particles that pass through this plane are considered uncollected. Assuming a nearly perfect elastic collision between the solid particles and cyclone walls, the coefficient of restitution in tangential as well as wall-normal direction is taken as unity.

Two criteria were chosen for convergence viz. the residuals and the monitored flow variables. Regarding the former, the simulation was considered converged when the flow residuals fall below the targeted value of 10^{-5}, and for the latter, the tangential velocity was monitored at several locations in the flow domain and it was ensured that the monitored value becomes (quasi-) steady with time.

2.4. Validation

For validation, we use the experimental data of Slack et al. [54] in which the Stairmand [46] with main body diameter, D = 0.205 m was used. The geometrical details have already been presented in Table 1, with the ST2 variant resembling the one used in [54]. At the inlet, a uniform velocity profile has been prescribed amounting to $Re = 2.58 \times 10^5$, as used in the experiment.

Figure 4 elucidates the comparison of the present numerical simulations against the experimental data by [54]. The first column represents radial profiles of mean axial velocity, and the second column shows the mean tangential velocity at axial locations Z/D = 0.811, 1.104, 2.128, and 3.006 (from top to bottom), respectively. At all the axial locations, the mean axial velocity agrees well both qualitatively and quantitatively—the local dip near the cyclone axis and the local maxima on either side of the axis are very well captured. The tangential velocity is also in good agreement with the reference data. Although a mild difference is observed in the upper part of the cyclone, in the conical section the local peaks match very well. Since the cyclonic flows are known to constitute a high level of anisotropy followed by the Rankine vortex, the numerical results presented here may be considered to fit well with the measured data. Two levels of mesh consisting of nearly 385K and 624K hexahedra were tested. Since no appreciable difference was noticed, the results of level 2 have been presented in Figure 4.

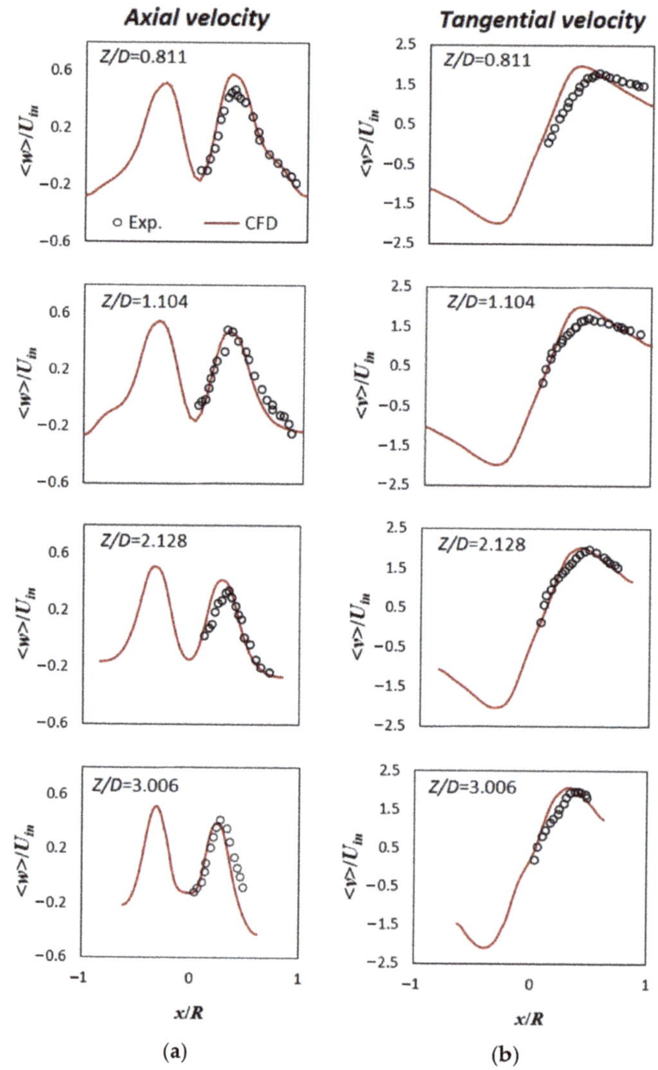

Figure 4. Comparative analysis of CFD results with experimental data [54]: (**a**) the first column—mean axial velocity; the second column—mean tangential velocity, and (**b**) from top to bottom: radial profiles at axial locations Z/D = 0.811, 1.104, 2.128, and 3.006 ($D = 2R$ = 0.205 m is the barrel diameter) from the cyclone roof, respectively.

A mild difference is observed in the axial velocity profiles and tangential velocity profiles when compared to the reference data due to a high level of anisotropy and complex flow phenomenon. Such differences are also reported in several studies even with the advanced closure LES; for instance [1,5–10,12–20].

3. Results and Discussion

In this section, we present the performance parameters of each cyclone model, followed by a detailed analysis of the mean flow field. Contour plots are represented over the $Y = 0$ plane, whereas the mean radial profiles are calculated at different axial locations. Finally, the pressure drop, cut-off particle size and overall separation efficiency in all models—also at different U_{in}—are computed and compared.

3.1. Continuous Phase

The first row in Figure 5 shows the contour plot of mean static pressure over the $Y = 0$ plane (normalized with the density times the square of the inlet velocity magnitude). The pressure distribution is nearly the same in all the models. In the core region, a negative pressure zone is observed—this means that the pressure there falls below the atmospheric pressure (the gauge pressure values have been used). The static pressure near the (outer) solid walls in the separation region is the highest, and this value reduces significantly when approaching the cyclone centre. The first row in Figure 6 illustrates the radial profiles of mean static pressure at $z/D = 1.0$, 2.0, and 3.0, from left to right, respectively. When moving along a radial direction, the pressure variations are smaller in the outer vortex region, and significant in the core region—large pressure gradients exist in the latter zone. The wall static pressure is the largest in the CC variant and lowest in the CV variant, with intermediate values in the ST variant. Furthermore, in each variant, the pressure values reduce from model 1 to 3. The wall static pressure reduces largely near the cyclone bottom, particularly in the CC variant, and in model ST1 (the same can be observed at the lowest axial location ($z/D = 3.0$)). It is also interesting to notice that the wall pressure is highly asymmetric in cyclone model CC1. This is due to a significant bending of the vortex core and a reduction in the cross-sectional area down the cyclone. The cyclone models having larger wall pressure values are expected to yield larger pressure drops.

The second row in Figure 5 presents the contour plots of mean axial velocity (normalized with the inlet velocity). It becomes apparent that this velocity component has two regions viz. the outer vortex swirling in the downward direction and the inner vortex directed upward—the former takes up the negative axial velocity values, whereas the latter is positive. Seemingly, the axial velocity pattern is nearly similar in all variants, except that the zone confined to the upward movement of the fluid is narrower in the CC variant and wider in the CV variant. The second row in Figure 6 represents the mean radial profiles at different locations. The downward directed flow has maximum magnitude near the outer walls, and it reduces when moving radially inwards till it becomes zero at a radial location. Thereafter, the axial velocity takes positive values in the rest of the region, indicating a movement in the upward direction. Furthermore, the axial velocity exhibits a local dip near the cyclone axis that takes up a small negative value at the geometric axis, indicating a narrow backflow region. The variations are more in the outer region than in the core region—at $z/D = 1.0$, small changes persist, and this difference increases with z/D values. At $z/D = 2.0$, larger variations can be noticed near the walls and become maximum at $z/D = 3.0$. In the CC variant, the axial velocity near the walls is much higher, particularly in the lower sections of the cyclone. The variations are significantly high with changes in the length of the cylindrical section in each model, particularly in the cone region. Also Demir et al. [55] noted a significant influence of the length of the cylindrical section on the values of the velocity components.

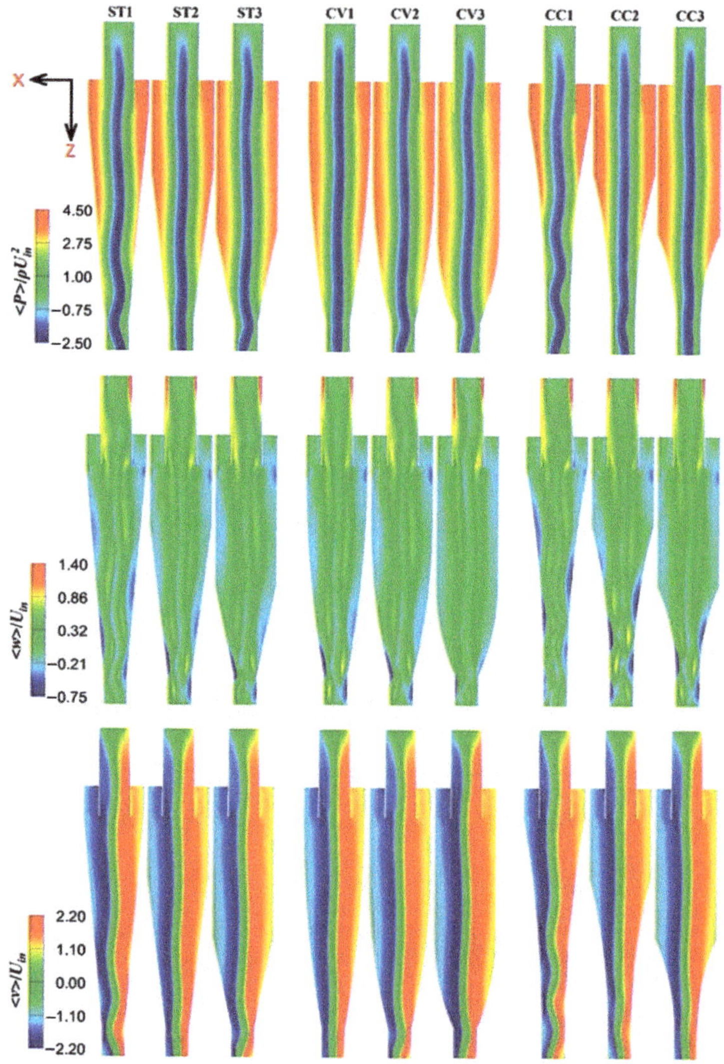

Figure 5. Contour plots on $Y/D = 0$ plane at $U_{in} = 10$ m/s. From left to right: different cyclone models. From top to bottom: mean static pressure, mean axial velocity, and mean tangential velocity.

The third row in Figure 5 elucidates the contour plots of mean tangential velocity in all models. This is the most important component of velocity as it directly controls the strength of the centrifugal force field generated in cyclones, which in turn influences the collection efficiency and pressure drop. The tangential velocity distribution is nearly the same in all cyclone models, and it is seemingly zero near the geometric axis; thereafter it takes up the maximum value in the core region. More detailed information can be found in the last line of Figure 6 which shows the radial profiles of mean tangential velocity at different axial locations. It becomes apparent that in the core region, the tangential velocity varies linearly—it is zero near the rotational axis of the inner vortex and increases linearly in the core region till it reaches the maximum value. Thereafter, the tangential velocity magnitude decreases rapidly in a nonlinear manner and reduces significantly near the outer walls. The peak value of tangential velocity is maximum in the CC variant and minimum in the CV variant. Furthermore, in each variant, the velocity decreases from model 1 to 3. Interestingly, the outer vortex region is very narrow in the CC variant, particularly in the lower part of the cyclone where a large asymmetric distribution is a sough. The region confined to the inner vortex is hardly affected. Since tangential velocity governs the centrifugal flow field, the cyclone models experiencing larger tangential magnitude are likely to undergo larger pressure drops.

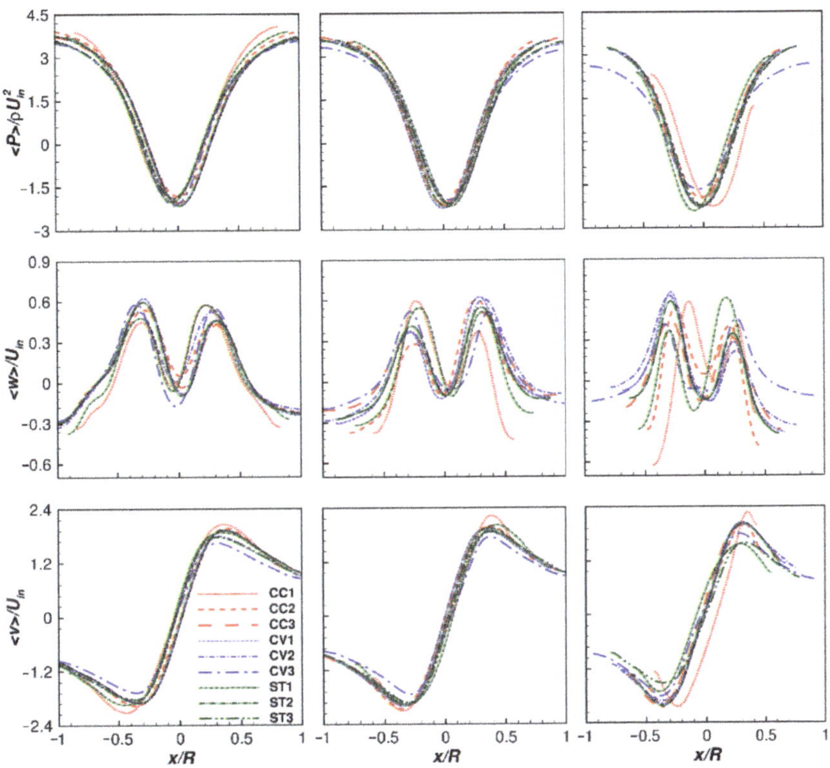

Figure 6. From left to right: the radial profiles at z/D = 1.0, 2.0, and 3.0 of mean static pressure, mean axial velocity and mean tangential velocity (top to bottom), respectively, in different cyclone models.

3.2. Pressure Drop

In the present study, the pressure drop is taken as the difference in the mean total pressure values at the inlet surface and outlet surface. This factor is proportional to the energy that a cyclone would require to operate. Higher pressure drop signifies large irreversible losses, and so, higher energy is required to make the fluid flow through the cyclone volume, and vice-versa. Since these devices operate continuously for a longer time in several applications, the proposed design must ensure low-pressure losses.

Figure 7 presents the pressure drop values in the cyclone variants at different inlet velocities, categorized as variants. With an increase in the inlet velocity, pressure drop increases significantly. For a given cylinder length, in comparison with the ST variant, the pressure drop is lowest in models 1 and 2 of the CV variant, whereas in the CC variant all the models elucidate an increased pressure drop. At U_{in} = 10 m/s and 15 m/s, models 1 and 2 in the CV variant shows lower pressure losses than in the ST variant. Interestingly, for all U_{in}, the pressure drop value in model 3 of the CV variant is greater than in the ST variant (as an exception, a similar observation also holds for the CV1 model at U_{in} = 20 m/s). This could be due to the larger cross-sectional area just above the downcomer tube in the CV variant than in ST and CC variants—the area is the largest in model 3. Due to this, the flow experiences significant contraction over a small height (above the top connecting surface of the downcomer tube) that complicates the flow, particularly at higher inlet velocities. In contrast to the CV variant, the results are more consistent in the CC variant when compared with the ST variant. Here, at a given U_{in} and cylinder length, the pressure drop is larger in all the models in the CV variant than in the CC variant. Another important observation is that the pressure drop also reduces with an increase in the length of the cylindrical section at a given U_{in}. A similar trend was found by Demir et al. [55] in their studies on effects of cylindrical and conical heights on pressure and velocity fields. Table 2 elucidates complete information on (percent change in) pressure drop with respect to the base variant model ST2, at different U_{in}, in all models.

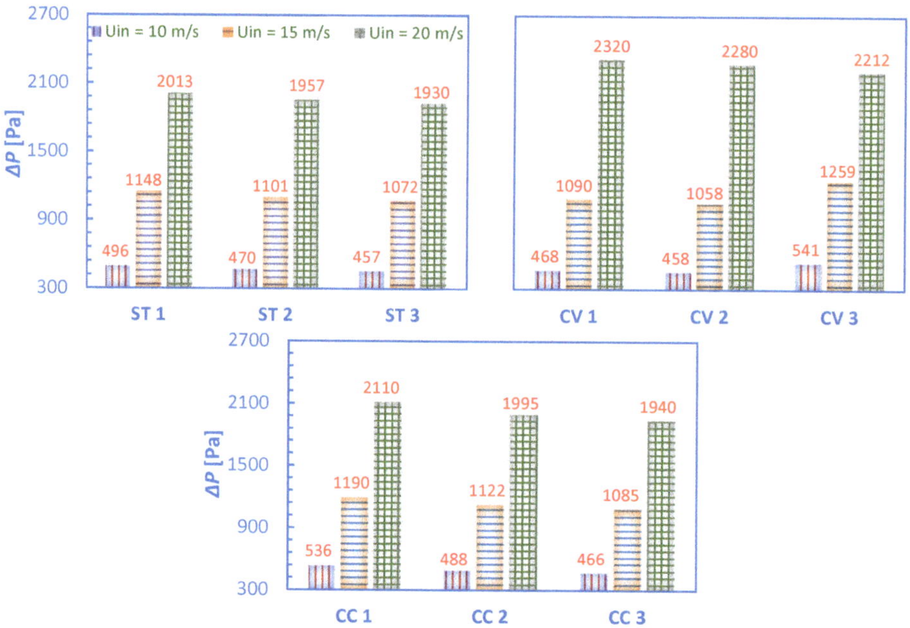

Figure 7. Pressure drop values in all the cyclone models of ST variant, CV variant and CC variant at inlet velocities, U_{in} = 10, 15 and 20 m/s.

Table 2. Pressure drop at different inlet velocities.

Cyclone Variants	Cyclone Models	% Change in Pressure Drop *		
		U_{in} = 10 m/s	U_{in} = 15 m/s	U_{in} = 20 m/s
ST	ST1	5.53	4.27	2.86
	ST2	–	–	–
	ST3	−2.77	−2.63	−1.38
CV	CV1	−0.425	−1.00	18.55
	CV2	−2.55	−3.91	16.51
	CV3	15.11	14.35	22.12
CC	CC1	14.04	8.08	7.82
	CC2	3.83	1.91	1.94
	CC3	−0.85	−1.45	−0.87

* % Change w.r.t. the standard cyclone model ST2.

3.3. Analysis of the Separation Efficiency

The main purpose of utilizing a cyclone is to separate the particles from gas, and in this regard, the separation capability of a cyclone is defined in two ways. First, the cut-off particle size (designated as x_{50})—defined as the particle size that corresponds to 50% separation efficiency on the grade efficiency curve (GEC). Second, the overall separation efficiency is calculated as the ratio of the number of particles separated by the cyclone separator to the number of particles entering the cyclone.

Figure 8 represents the GEC of all variants at U_{in} = 10 m/s. The change in dimensions and shape of individual sections of the cyclone separator did not significantly change the profile of GEC curves. Similar observations were made by Shastri et al. [39] and Xiang et al. [37]. For all cylinder lengths in each variant, the variations in the GECs are from mild to moderate—CC variants are more sensitive to the cylinder height than the ST variant, whereas in the CV variant the variations are very mild. Complete information on the cut-off size for all U_{in} is presented in Figure 9. Two important observations here are: with an increase in cylinder length, x_{50} increases mildly at all U_{in}, whereas with an increase in U_{in}, the cut-off size reduces significantly in all variants. For short cylinder height and lower U_{in}, the CC variant outperforms the ST as well as CV variants, whereas, at higher U_{in}, the x_{50} values in the ST variant are greater than other variants by a very small margin.

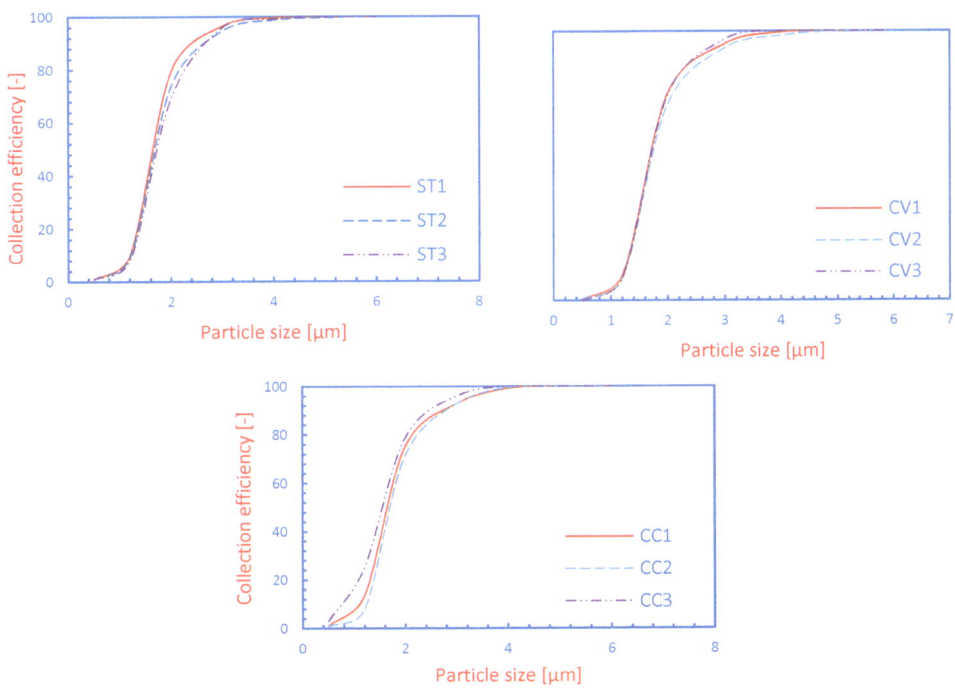

Figure 8. Grade efficiency curves for all cyclone models in ST variant, CV variant and CC variant at inlet velocities U_{in} = 10 m/s with particle sizes ranging from 0.4 to 8.0 µm.

Figure 10 illustrates the overall separation efficiency in all variants at different U_{in}. With an increase in U_{in}, the overall separation efficiency in each variant increases significantly, whereas the increase in the cylinder height mildly reduces the collection efficiency at a given U_{in}. The efficiency of different models in the CC variant is slightly better than the standard ST variant, and in the case of CV, the efficiency is the lowest.

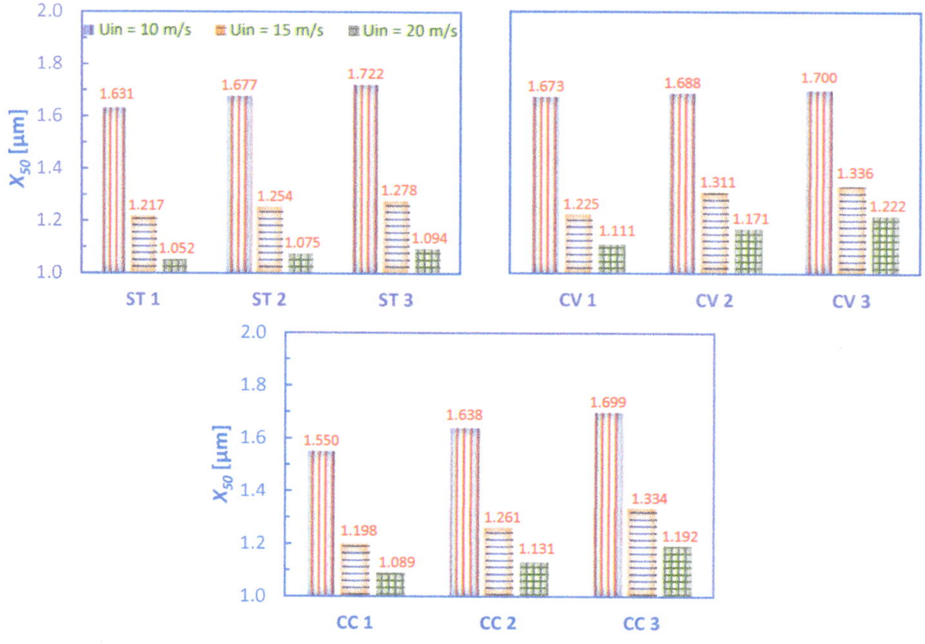

Figure 9. Cut-off sizes in all cyclone models of ST variant, CV variant, and CC variant at inlet velocities U_{in} = 10, 15 and 20 m/s.

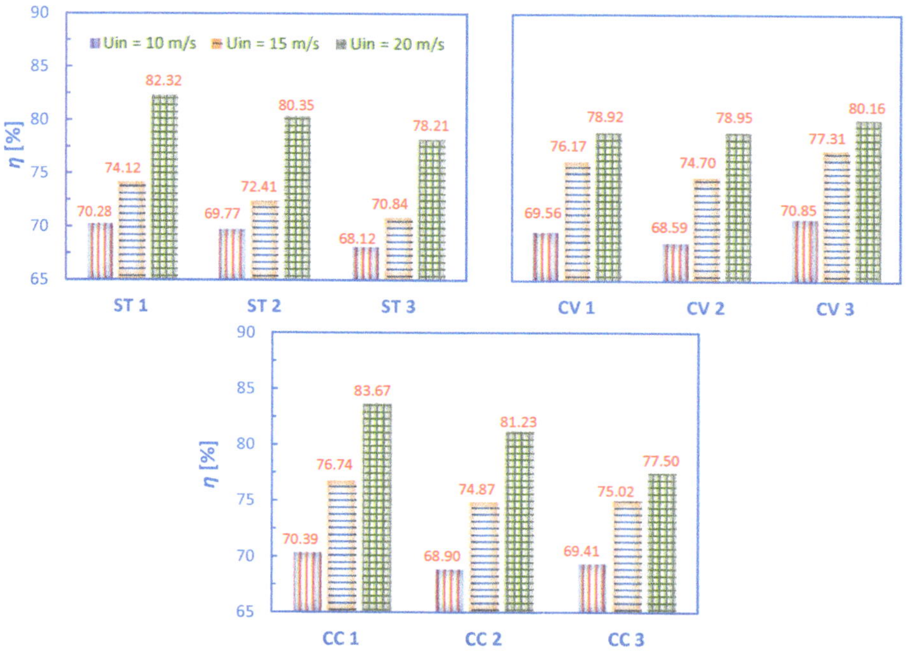

Figure 10. Overall separation efficiency in all the cyclone models of ST variant, CV variant and CC variant at inlet velocities U_{in} = 10, 15 and 20 m/s.

Table 3 provides complete information on (percent change in) overall separation efficiency and cut-off size with respect to the base variant ST2, at different U_{in}, in all variants.

Table 3. Percent change in overall separation efficiency and cut-off diameter at different U_{in}.

Cyclone Variants	Cyclone Models	% Change in Overall Separation Efficiency *			% Change in Cut-Off Diameter *		
		U_{in} = 10 m/s	U_{in} = 15 m/s	U_{in} = 20 m/s	U_{in} = 10 m/s	U_{in} = 15 m/s	U_{in} = 20 m/s
ST	ST1	0.73	2.36	2.45	−2.16	−2.95	−2.14
	ST2	–	–	–	–	–	–
	ST3	−2.37	−2.17	−2.66	3.30	1.91	1.77
CV	CV1	−0.31	5.19	−1.78	0.36	−2.31	3.35
	CV2	−1.69	3.16	−1.74	1.26	4.55	8.93
	CV3	1.55	6.77	−0.24	1.98	6.54	13.67
CC	CC1	0.89	5.98	4.13	−7.02	−4.47	1.30
	CC2	−1.25	3.40	1.10	−1.74	0.56	5.21
	CC3	−0.52	3.60	−3.55	1.92	6.38	10.88

* % Change w.r.t. the standard cyclone model ST2.

4. Conclusions

In the present study, unsteady simulations were performed using RSM as a closure model to RANS. The airflow was simulated over the stationary grids, whereas the solid particles were tracked using the Lagrangian approach. Hence, we made use of the Eulerian-Lagrangian approach to model the cyclonic flow field. Nine cyclone models were simulated at three different inlet velocities viz. U_{in} = 10 m/s, 15 m/s, and 20 m/s—thus, a total of twenty-seven cases were simulated. We considered three cyclone variants viz. a straight wall of the conical section as in the conventional cyclone design (referred to as ST variant), a curved wall with convex profile (referred to as CV variant), and a curved wall with concave profile (referred to as CC variant)—the convex and concave terminology was adopted with reference to an observer outside the cyclone structure. Furthermore, three different lengths of the cylindrical section were considered in all the three cyclone variants keeping the total cyclone length a constant.

Based on the analysis of the obtained results, the following conclusions were defined:

- With the increase in inlet velocity:
  a. Pressure drop increased significantly in all cyclone variants.
  b. Cut-off particle size reduced tremendously.
  c. Collection efficiency increased by a large amount.
- With an increase in the length of the cylindrical section:
  a. Pressure drop was reduced in all the cyclone variants.
  b. Cut-off particle size increased mildly.
  c. Collection efficiency reduced by a marginal amount.
- With wall profile:
  a. Pressure drop was the lowest in models 1 and 2 of the CV, particularly at U_{in} = 10 m/s, and 15 m/s. At U_{in} = 20 m/s, pressure drop was the largest in CV variants, and at U_{in} = 10 m/s, and 15 m/s, as compared with the ST variant, the variations were moderate.
  b. Cut-off particle size showed a mild variation at a given U_{in}. The cut-off size was largest in CV variants, whereas this size was comparable in ST and CC variants.
  c. Collection efficiency increased by a marginal amount in the CC variant and was minimum in the CV variant.

Author Contributions: Conceptualization, L.S.B.; methodology, L.S.B. and M.W.; software, S.P., I.S., O.P., T.M., L.S.B. and M.W.; validation, S.P. and M.W.; formal analysis, M.W.; investigation, S.P., L.S.B. and M.W.; resources, S.P., I.S., O.P. and T.M.; data curation, S.P., L.S.B., M.W., J.I. and A.K.R.; writing—original draft preparation, S.P., L.S.B. and M.W.; writing—review and editing, S.P., L.S.B., M.W., L.S.B., J.I. and A.K.R.; visualization, S.P., L.S.B. and M.W.; supervision, L.S.B., and M.W.; project administration, L.S.B. All authors have read and agreed to the published version of the manuscript.

Funding: This research received no external funding.

Institutional Review Board Statement: Not applicable.

Informed Consent Statement: Not applicable.

Data Availability Statement: Not applicable.

Conflicts of Interest: The authors declare no conflict of interest.

References

1. Brar, L.S.; Derksen, J.J. Revealing the details of vortex core precession in cyclones by means of large-eddy simulation. *Chem. Eng. Res. Des.* **2020**, *159*, 339–352. [CrossRef]
2. Faulkner, W.B.; Shaw, B.W. Efficiency and pressure drop of cyclones across a range of inlet velocities. *Appl. Eng. Agric.* **2006**, *22*, 155–161. [CrossRef]
3. Safikhani, H.; Akhavan-Behabadi, M.A.; Shams, M.; Rahimyan, M.H. Numerical simulation of flow field in three types of standard cyclone separators. *Adv. Powder Technol.* **2010**, *21*, 435–442. [CrossRef]
4. Hoffman, A.C.; Stein, L.E. *Gas Cyclones and Swirl Tubes: Principles, Design and Operation*; Springer: Berlin/Heidelberg, Germany, 2008. [CrossRef]
5. Brar, L.S.; Sharma, R.P.; Dwivedi, R. Effect of vortex finder diameter on flow field and collection efficiency of cyclone separators. *Part. Sci. Technol.* **2014**, *33*, 34–40. [CrossRef]
6. Wasilewski, M.; Brar, L.S. Optimization of the geometry of cyclone separators used in clinker burning process: A case study. *Powder Technol.* **2017**, *313*, 293–302. [CrossRef]
7. Wasilewski, M.; Brar, L.S.; Ligus, G. Experimental and numerical investigation on the performance of square cyclones with different vortex finder configurations. *Sep. Purif. Technol.* **2020**, *239*, 1–20. [CrossRef]
8. Elsayed, K.; Lacor, C. The effect of cyclone vortex finder dimensions on the flow pattern and performance using LES. *Comput. Fluids* **2013**, *71*, 224–239. [CrossRef]
9. Parvaz, F.; Hosseini, S.H.; Ahmadi, G.; Elsayed, K. Impacts of the vortex finder eccentricity on the flow pattern and performance of a gas cyclone. *Sep. Purif. Technol.* **2017**, *187*, 1–13. [CrossRef]
10. Brar, L.S.; Elsayed, K. Analysis and optimization of cyclone separators with eccentric vortex finders using large eddy simulation and artificial neural network. *Sep. Purif. Technol.* **2018**, *207*, 269–283. [CrossRef]
11. Pei, B.; Yang, L.; Dong, K.; Jiang, Y.; Du, X.; Wang, B. The effect of cross-shaped vortex finder on the performance of cyclone separator. *Powder Technol.* **2017**, *313*, 135–144. [CrossRef]
12. Misiulia, D.; Antonyuk, S.; Andersson, A.G.; Lundström, T.S. Effects of deswirler position and its centre body shape as well as vortex finder extension downstream on cyclone performance. *Powder Technol.* **2018**, *336*, 45–56. [CrossRef]
13. Kumar, V.; Jha, K. Multi-objective shape optimization of vortex finders in cyclone separators using response surface methodology and genetic algorithms. *Sep. Purif. Technol.* **2019**, *215*, 25–31. [CrossRef]
14. Brar, L.S.; Sharma, R.P. Effect of varying diameter on the performance of industrial scale gas cyclone dust separators. *Mater. Today Proc.* **2015**, *4*, 3230–3237. [CrossRef]
15. Brar, L.S. Application of response surface methodology to optimize the performance of cyclone separator using mathematical models and CFD simulations. *Mater. Today Proc.* **2018**, *5*, 20426–20436. [CrossRef]
16. Brar, L.S.; Sharma, R.P.; Elsayed, K. The effect of the cyclone length on the performance of Stairmand high-efficiency cyclone. *Powder Technol.* **2015**, *286*, 668–677. [CrossRef]
17. Elsayed, K.; Lacor, C. The effect of cyclone inlet dimensions on the flow pattern and performance. *Appl. Math. Model.* **2011**, *35*, 1952–1968. [CrossRef]
18. Wasilewski, M.; Brar, L.S. Effect of the inlet duct angle on the performance of cyclone separators. *Sep. Purif. Technol.* **2019**, *213*, 19–33. [CrossRef]
19. Brar, L.S.; Elsayed, K. Analysis and optimization of multi-inlet gas cyclones using large eddy simulation and artificial neural network. *Powder Technol.* **2017**, *311*, 465–483. [CrossRef]
20. Misiulia, D.; Andersson, A.G.; Lundström, T.S. Effects of the inlet angle on the collection efficiency of a cyclone with helical-roof inlet. *Powder Technol.* **2017**, *305*, 48–55. [CrossRef]
21. Piemjaiswang, R.; Ratanathammapan, K.; Kunchonthara, P.; Piumsomboon, P.; Chalermsinsuwan, B. CFD study of cyclone performance: Effect of inlet section angle and particle size distribution. *J. Teknol.* **2016**, *78*, 83–89. [CrossRef]

22. Yao, Y.; Huang, W.; Wu, Y.; Zhang, Y.; Zhang, M.; Yang, H.; Lyu, J. Effects of the inlet duct length on the flow field and performance of a cyclone separator with a contracted inlet duct. *Powder Technol.* **2021**, *393*, 12–22. [CrossRef]
23. Wang, Z.; Sun, G.; Jiao, Y. Experimental study of large-scale single and double inlet cyclone separators with two types of vortex finder. *Chem. Eng. Process.-Process Intensif.* **2020**, *158*, 108188. [CrossRef]
24. Qiang, L.; Qinggong, W.; Weiwei, X.; Zilin, Z.; Konghao, Z. Experimental and computational analysis of a cyclone separator with a novel vortex finder. *Powder Technol.* **2020**, *360*, 398–410. [CrossRef]
25. Le, D.K.; Yoon, J.Y. Numerical investigation on the performance and flow pattern of two novel innovative designs of four-inlet cyclone separator. *Chem. Eng. Process.-Process Intensif.* **2020**, *150*, 107867. [CrossRef]
26. Noh, S.Y.; Heo, J.E.; Woo, S.H.; Kim, S.J.; Ock, M.H.; Kim, Y.J.; Yook, S.J. Performance improvement of a cyclone separator using multiple subsidiary cyclones. *Powder Technol.* **2018**, *338*, 145–152. [CrossRef]
27. Wasilewski, M.; Brar, L.S.; Ligus, G. Effect of the central rod dimensions on the performance of cyclone separators-optimization study. *Sep. Purif. Technol.* **2021**, *274*, 119020. [CrossRef]
28. Parvaz, F.; Hosseini, S.H.; Elsayed, K.; Ahmadi, G. Influence of the dipleg shape on the performance of gas cyclones. *Sep. Purif. Technol.* **2020**, *233*, 116000. [CrossRef]
29. Xu, M.; Yang, L.; Sun, X.; Wang, J.; Gong, L. Numerical analysis of flow resistance reduction methods in cyclone separator. *J. Taiwan Inst. Chem. Eng.* **2019**, *96*, 419–430. [CrossRef]
30. Qiang, L.; Jianjun, W.; Weiwei, X.; Meng, Z. Investigation on separation performance and structural optimization of a two-stage series cyclone using CPFD and RSM. *Adv. Powder Technol.* **2020**, *31*, 3706–3714. [CrossRef]
31. Galletti, C.; Rum, A.; Turchi, V.; Nicolella, C. Numerical analysis of flow field and particle motion in a dynamic cyclonic selector. *Adv. Powder Technol.* **2020**, *31*, 1264–1273. [CrossRef]
32. Sun, Z.; Liang, L.; Liu, Q.; Yu, X. Effect of the particle injection position on the performance of a cyclonic gas solids classifier. *Adv. Powder Technol.* **2020**, *31*, 227–233. [CrossRef]
33. Fu, S.; Zhou, F.; Sun, G.; Yuan, H.; Zhu, J. Performance evaluation of industrial large-scale cyclone separator with novel vortex finder. *Adv. Powder Technol.* **2021**, *32*, 931–939. [CrossRef]
34. Mofarrah, M.; Hojjat, Y.; Mashayekh, S.; Liu, Z.; Yan, K. Introduction and simulation of a small electro cyclone for collecting indoor pollen particles. *Adv. Powder Technol.* **2021**, *33*, 103384. [CrossRef]
35. An, I.H.; Lee, C.H.; Lim, J.H.; Lee, H.Y.; Yook, S.J. Development of a miniature cyclone separator operating at low Reynolds numbers as a pre-separator for portable black carbon monitors. *Adv. Powder Technol.* **2021**, *32*, 4779–4787. [CrossRef]
36. Gimbun, J.; Chuah, T.G.; Choong, T.S.Y.; Fakhru'l-Razi, A. Prediction of the effects of cone tip diameter on the cyclone performance. *Aerosol Sci.* **2005**, *36*, 1056–1065. [CrossRef]
37. Xiang, R.; Park, S.H.; Lee, K.W. Effects of cone dimension on cyclone performance. *J. Aerosol Sci.* **2001**, *32*, 549–561. [CrossRef]
38. Stern, A.; Caplan, K.; Bush, P. *Cyclone Dust Collectors*; American Petroleum Institute: New York, NY, USA, 1955.
39. Shastri, R.; Brar, L.S. Numerical investigations of the flow-field inside cyclone separators with different cylinder-to-cone ratios using large-eddy simulation. *Sep. Purif. Technol.* **2020**, *249*, 1–17. [CrossRef]
40. Shastri, R.; Sharma, R.P.; Brar, L.S. Numerical investigations of cyclone separators with different cylinder-to-cone ratios. *Part. Sci. Technol.* **2021**, *40*, 337–345. [CrossRef]
41. Ghodrat, M.; Kuang, S.B.; Yu, A.B.; Vince, A.; Barnett, G.D.; Barnett, P.J. Numerical analysis of hydrocyclones with different conical section designs. *Miner. Eng.* **2014**, *62*, 74–84. [CrossRef]
42. Wilcox, D.C. *Turbulence Modeling for CFD*; DCW Industries, Inc.: La Cañada Flintridge, CA, USA, 1994.
43. *Fluent Theory Guide*; Fluent Inc.: Canonsburg, PA, USA, 2016.
44. Morsi, A.J.; Alexander, S.A. An Investigation of Particle Trajectories in Two-Phase Flow System. *J. Fluid Mech.* **1972**, *55*, 193–208. [CrossRef]
45. Shukla, S.K.; Shukla, P.; Ghosh, P. Evaluation of Numerical Schemes for Dispersed Phase Modeling of Cyclone Separators. *Eng. Appl. Comp. F Mech.* **2010**, *5*, 235–246. [CrossRef]
46. Stairmand, C.J. The design and performance of cyclone separators. *Ind. Eng. Chem.* **1951**, *29*, 356–383.
47. Wasilewski, M. Analysis of the effects of temperature and the share of solid and gas phases on the process of separation in a cyclone suspension preheater. *Sep. Purif. Technol.* **2016**, *168*, 114–123. [CrossRef]
48. Wasilewski, M. Analysis of the effect of counter-cone location on cyclone separator efficiency. *Sep. Purif. Technol.* **2017**, *179*, 236–247. [CrossRef]
49. Wasilewski, M.; Brar, L.S. Investigations of the flow field inside a square cyclone separator using DPIV and CFD. *E3S Web Conf.* **2019**, *100*, 1–8. [CrossRef]
50. Venkatesh, S.; Kumar, R.S.; Sivapirakasam, S.P.; Sakthivel, M.; Venkatesh, D.; Arafath, S.Y. Multi-objective optimization, experimental and CFD approach for performance analysis in a square cyclone separator. *Powder Technol.* **2020**, *371*, 115–129. [CrossRef]
51. Wasilewski, M.; Duda, J. Multicriteria optimisation of first-stage cyclones in the clinker burning system by means of numerical modelling and experimental research. *Powder Technol.* **2016**, *289*, 143–158. [CrossRef]
52. Stachnik, M.; Jakubowski, M. Multiphase model of flow and separation phases in a whirlpool: Advanced simulation and phenomena visualization approach. *J. Food Eng.* **2020**, *274*, 109846. [CrossRef]

53. Jakubowski, M.; Sterczyska, M.; Matysko, R.; Poreda, A. Simulation and experimental research on the flow inside a whirlpool separator. *J. Food Eng.* **2014**, *133*, 9–15. [CrossRef]
54. Slack, M.D.; Prasad, R.O.; Bakker, A.; Boysan, F.M. Advances in Cyclone Modelling Using Unstructured Grids. *Chem. Eng. Res. Des.* **2000**, *78*, 1098–1104. [CrossRef]
55. Demir, S.; Karadeniz, A.; Aksel, M. Effects of cylindrical and conical heights on pressure and velocity fields in cyclones. *Powder Technol.* **2016**, *295*, 209–217. [CrossRef]

applied sciences

Article

Estimating the Potential Differential Settlement of a Tailings Deposit Based on Consolidation Properties Heterogeneity

Mohammad Mahdi Badiozamani * and Nicholas Beier

Department of Civil and Environmental Engineering, University of Alberta, Edmonton, AB T6G 1H9, Canada; nabeier@ualberta.ca
* Correspondence: mbadiozaman@ualberta.ca

Abstract: Processing of extracted oil sands generates substantial volumes of tailings slurries. Due to the scale and inherent variability of the tailings properties, consolidation settlement is expected to occur at different rates and magnitudes across the tailings deposit. Estimating potential differential settlement of the consolidated deposit surface is an essential input for closure design. This paper presents a three-step methodology that generates multiple realizations of quasi-three-dimensional (3D) surfaces of the consolidated deposit based on the adjacent points. Each point is based on a stochastic one-dimensional (1D) large strain consolidation model developed with Monte Carlo techniques in GoldSim. The simulated surfaces provide early estimates of differential settlement based on the variability of consolidation properties expected in the tailings deposit. Comprehensive sensitivity analyses are performed for differently treated tailings material through 28 distinct scenarios to evaluate the sensitivity of the developed 1D and 3D models to consolidation input parameters over a 40-year time period. The analysis demonstrated that differential settlement is highly sensitive to tailings compressibility and hydraulic conductivity governed by the constitutive relationship parameters, and less sensitive to the solids content, specific gravity or thickness of a surcharge load. Tailings that underwent steady continuous settlement exhibited the largest degree of differential settlement.

Keywords: differential settlements; Monte-Carlo simulation; consolidation; heterogeneity; GoldSim; soil cover

Citation: Badiozamani, M.M.; Beier, N. Estimating the Potential Differential Settlement of a Tailings Deposit Based on Consolidation Properties Heterogeneity. *Appl. Sci.* **2022**, *12*, 6206. https://doi.org/10.3390/app12126206

Academic Editors: Hassane Naji, María Isabel Lamas Galdo and Rodriguez J. D.

Received: 20 May 2022
Accepted: 16 June 2022
Published: 18 June 2022

Publisher's Note: MDPI stays neutral with regard to jurisdictional claims in published maps and institutional affiliations.

Copyright: © 2022 by the authors. Licensee MDPI, Basel, Switzerland. This article is an open access article distributed under the terms and conditions of the Creative Commons Attribution (CC BY) license (https://creativecommons.org/licenses/by/4.0/).

1. Introduction

The Athabasca Oil Sands Region in Alberta, Canada, is home to the third-largest oil reserves after Venezuela and Saudi Arabia. As of 2016, Alberta's oil sands proven reserves were 165.4 billion barrels. Surface mining recovers 20% of this total volume and covers an area of about 4800 km^2 [1]. Fluid fine tailings (FFT) are a by-product of surface bitumen extraction activities, which have accumulated a total volume of 1.3 billion m^3 [2].

The government of Alberta views oil sands operations as temporary land use and requires oil sands operators to return disturbed land to equivalent land capability [3]. Reclamation efforts for such large-scale disturbances require the reconstruction of new landscape-scale ecosystems [4,5]. Revegetation is key to reclamation in a terrestrial landscape. Therefore, reclamation soil covers are designed to provide an adequate water supply for vegetation over dry summer periods [6,7]. However, most reclamation soil cover research does not explicitly consider the 3D nature of differential settlement within the context of large-scale landscapes, such as oil sand tailings deposits.

Understanding the dynamic nature in which reclamation takes place is fundamental in re-establishing ecosystem functionality. One of the constraints in landscape design is to limit the free water area to less than 10–20% of the reclaimed deposit [8,9]. However, differential settlement may result in shallow ponded water and wetlands or may deepen or expand the existing wet landscape features. There is a wealth of hydrological studies that quantify downslope movement of water within reclaimed upland slopes and temporal variability of soil water distribution [10–12]. However, little effort seems to be directed at the spatial variability of underlying tailings with respect to geotechnical stability and settlement. The large strain consolidation theory [13] is commonly used to model the non-linear behavior of oil sand tailings consolidation. The spatial variability is of significance when reclamation covers overlay such highly compressible FFT. There is certainly a need to understand not only the settlement with time for these covers but also any potential 3D surface roughness or differential settlement, as it directly affects the contour of the final surface. In addition to reclamation and landscape designing, analyzing the consolidation of the treated tailings is also an essential input to predict the storage capacity of tailings storage facilities.

In any tailings storage facility, the three main stages involved in settlements of fine-grained tailings material are flocculation, sedimentation and consolidation [14]. As the consolidation stage dominates the long-term settlement, the focus is typically on analyzing the consolidation stage [15]. A one-dimensional (1D) column consolidation is the basis for most of the current modeling practices and has been extrapolated to the whole deposits' consolidation behavior [16–18]. Another example of approximating the 3D consolidation based on a 1D large strain model is presented in [19], which studied the 3D analysis of a large strain consolidation model for tailings that are deposited in a mining pit. The goal was to estimate the filling time and tailings storage capacity. The authors performed Hydraulic Consolidation Test (HCT) to obtain the consolidation constitutive relationships, and generated approximate solutions of the 3D analysis. The proposed algorithm discretizes the pit volume in columns and layers to render upper and lower bound solutions based on a 1D large strain model.

The quasi-3D models have been developed to estimate the consolidation of tailings deposits, most of them based on a series of 1D columns of concentric cylinders or adjacent columns [20–22]. There exists a critical assumption in quasi-3D consolidation models; at the same elevation, the sides of the tailings storage facility face the same deformation as the tailings material faces in the center of the deposit. This assumption introduces estimation error to such 3D models, which is a function of the geometry of the tailings storage facility, boundary conditions, tailings material properties and the filling rate. For the tailings facilities with irregular impoundment geometry, non-homogeneous distribution of tailings and complex drainage paths, the error of quasi-3D consolidation models are magnified [23]. Due to these potentially large errors related to compressible foundation boundaries, the 3D large strain analyses are preferred over 1D large strain analysis for feasibility, and detailed level engineering work [24,25] developed an approach to eliminate the calculation errors caused by compressible boundaries of the tailings storage facilities. Reference [24] briefly discusses the common methods that are used in tailings consolidation modeling, with their relative accuracy and pitfalls, and specific recommendations and the applicability of each method based on the facility geometry, tailings properties and filling rates.

There are research works that address the multi-dimensional consolidation modeling without approximations based on 1D consolidation models. However, there are numerical challenges in multi-dimensional modeling of tailings consolidation. References [23,26] introduced a geotechnical numerical modeling approach to simulate the gradual tailings impoundment and its consolidation in multi-dimensional space based on the large strain consolidation theory. The proposed algorithm is based on the Norwegian Geotechnical Institute [NGI] modeling approach, and is developed on FLAC and FLAC3D, the commercially available specialized geotechnical modeling software. The authors validated the proposed algorithm using the benchmarks published in [27]. To model the gradual deposition of tailings, the slurry is divided into many continuous layers that are activated sequentially from the bottom. Once each new layer is activated, a large-strain consolidation stage is triggered and the consolidation time will be a function of discharge history and the layer volume. Published case studies were not found with statistical modeling capability to consider for heterogeneity and uncertainty of consolidating tailings deposits.

Reference [28] listed some of the effective techniques to measure the actual settlement of the field, including both remote sensing methods (photogrammetry, Light Detection and Ranging-Lidar) and on-tailings methods (hovercrafts and swamp machine). These post-deposition techniques provide a high accuracy of elevation data that is essential to generate a 3D model of tailings facilities for detailed design. The other group of techniques to generate a 3D model of the deposit surface includes different estimations of the potential surface profile using simulation techniques and numerical methods such as large strain consolidation theory [13]. These techniques work well as the tailings deposit can be fully characterized post deposition. However, successful closure planning begins during the early stages of the mine operations, before tailings are deposited or the mine operations break ground. Given the limited actual field deposit data described above that are available at the early stages of mine planning and operations, the impact of the variability of tailings properties within the deposit may not be fully appreciated. The merit of this paper is to generate a very first estimate of the surface roughness and differential settlements well before the tailings deposit is filled and construction of a cover begins.

The objective of this paper is to develop a framework for estimating the differential settlements of a consolidating tailings surface based on a series of adjacent consolidating 1D columns. This has been done through implementing a large strain consolidation model developed by [29,30]. The performance and utility of the developed model in this paper is demonstrated through several scenarios using a synthetic tailings deposit dataset. The sensitivity of numerical settlement predictions to solids contents, specific gravity and tailings constitutive relationships are also presented.

Different tailings types were investigated in this study, including composite tailings (CT), thickened tailings (TT), centrifuge cake and cyclone overflow. The presented framework is a tool to generate a first estimate of the probable surface roughness of a consolidating tailings deposit based on spatial heterogeneity in consolidation properties (including compressibility and hydraulic conductivity). As such, it is not aimed at estimating tailings storage volume. The developed model provides insight into the probable differential settlement behavior of a tailings deposit surface a priori, based on the anticipated range of variation in tailings properties.

2. Methodology

A 1D large strain consolidation model based on Gibson's theory was developed by [29,30], which is the primary input to the quasi-3D tailings surface contour model in this paper. Zheng's model simulates the consolidation of a column of deposited tailings and generates a height interface of the column over time, using System Dynamics techniques in the GoldSim environment.

The rationale to include uncertainty in the modeling is to consider the variability of tailings properties in a large deposit. Sources of variability in oil sands tailings properties include bitumen content, mineral solids and clay content, which is evident both spatially and temporally in tailings deposits, as reported by [31]. The tailings consolidation and geotechnical behavior are significantly influenced by the amount and type of clay present in the tailings [32–35], while both sand and clay contents are variable with depth of the deposit. This variability results in considerable heterogeneity in type and amount of clay minerals and the particle size distribution (PSD) of the tailings profile [36–38], which consequently influences the compressibility and permeability of the consolidated deposit [20,39]. The developed methodology captures the inherent uncertainty of tailings consolidation through multiple replications of the model and sampling from tailings' properties distributions (i.e., solids content, specific gravity and large strain constitutive relationships).

To estimate the surface roughness of the consolidating FFT, three steps are followed in a sequence, as illustrated in Figure 1. This methodology generates multiple quasi-3D profiles of the probable final surface from the 1D column consolidation GoldSim model. In the first step, the 1D consolidation model is run for multiple realizations through changing the solids content, specific gravity and sampling for the parameters of the consolidation constitutive relationships from user-defined probability distribution functions. The height interface curves are generated at the end of step one. A settlement curve is randomly selected in the second step and assigned to 'Defined points' in a two-dimensional (2D) grid. The Defined points are the points' locations where the deposit profile is assumed, and the value of solids contents, specific gravity, fines contents and the settlement curves that govern the settlement rate over time are therefore assigned accordingly. In deposits that have been filled, the 'Defined points' are the locations for which the characteristics such as solid contents and specific gravity are known as a result of field samplings. In step three, the elevation of all the remaining points of the surface, i.e., the 'Intermediate points', are calculated based on the inverse distance weighted averaging of the Defined points. As a result of step three, the surface at each time step is estimated, and if the model is run for multiple realizations, multiple potential surfaces will be generated. The three steps are performed in two different, but related GoldSim models.

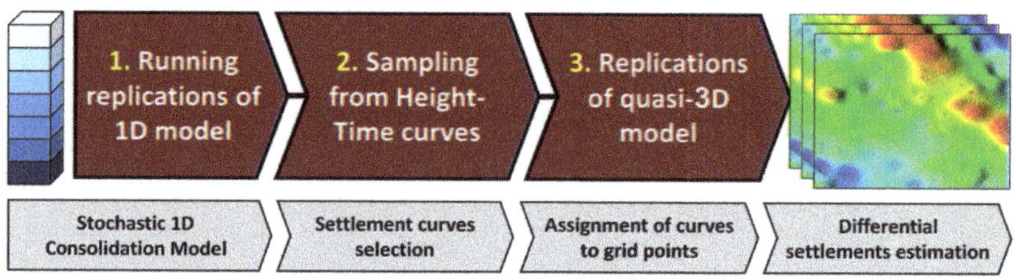

Figure 1. The methodology of quasi-3D consolidation modeling.

The methodology is implemented through the following steps: the settlement curves generated by the 1D GoldSim model are stored in a data file. A second data file contains all the inputs required to run the quasi-3D model, including the X and Y coordinates of all the grid points in separate tables for the Defined and Intermediate points. The quasi-3D GoldSim model reads these two files, calculates the elevation of Intermediate points over time using inverse distance weighted (IDW) averaging and writes them automatically in the output text file. At the final step, a series of Matlab functions read the text results, analyze the differential settlement and visualize the quasi-3D surface profiles and calculate the surface roughness at each specified time step and replication.

2.1. The 1D Model: Large Strain Consolidation Theory

The 1D large strain consolidation model developed by [30] is the basis for the developed methodology of this paper. The 1D model implements a finite difference scheme [40,41] to solve the large strain model with Gibson theory [13] using System Dynamics (SD) techniques. It simulates the long-term soil water dynamics and self-weight consolidation process using the Monte Carlo technique in GoldSim environment [42]. The GoldSim model is a reproducible open-source and transparent model that enables the technical and non-technical stakeholders to understand the mechanism of the simulation model. Reference [29] explicitly models two types of uncertainties: the aleatory uncertainties that are due to inherent randomness of the parameters, and the epistemic uncertainties that arise because of the lack of knowledge about different aspects of the problem.

Equations (1) and (2) present the constitutive relationships in the 1D model [43] and govern the change in the height, void ratio and effective stress of each given column of tailings material over time:

$$e = A \times \sigma'^{B} \tag{1}$$

$$k = C \times e^{D} \tag{2}$$

where e is the void ratio, σ' is the effective stress, k is the hydraulic conductivity and A, B, C and D are the constitutive relationships parameters. These parameters are determined from curve fitting of experimental data and are unique to each tailings sample.

2.2. Inverse Distance Weighted (IDW) Averaging

The elevation of Intermediate points in the surface at any given time step is estimated based on the weighted average inverse distance calculation in GoldSim, as presented in Equation (3):

$$Z_x = \sum_{i=1}^{S} \frac{Z_i}{(d_i)^m} / \sum_{i=1}^{S} \frac{1}{(d_i)^m} \tag{3}$$

where, d_i is the Euclidean distance between point x (the point of interest) and point i in the XY plane, S is the number of all Defined points, m is the inverse distance order and Z_x is the elevation of Intermediate points on the surface at any given time. The formulation simply implies that the elevation of Intermediate point x is the average of the elevation of all Defined points. The closer the Defined point to x, the greater impact it has on Z_x.

Parameter m quantifies the significance of closeness of Defined points to point x in determining Z_x. Choosing higher values, e.g., $m = 3$ (cubed inverse distance) means that as the distance increases, the contribution of the farther points' elevation in the average is significantly lower than that of closer points, compared to the case where $m = 1$ (simple inverse distance). An example is illustrated in Figure 2 to show how parameter m affects the quasi-3D estimation result.

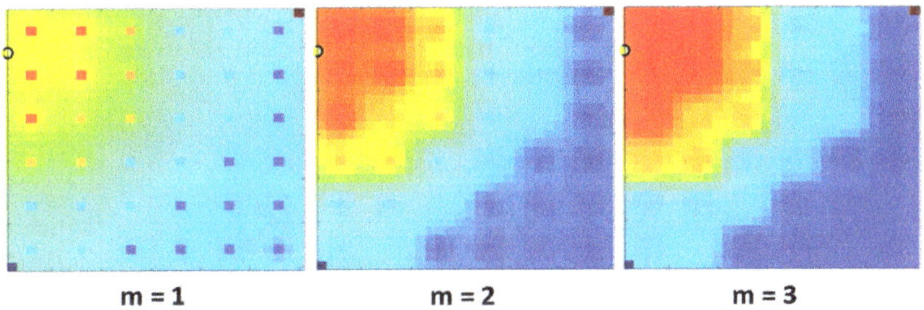

Figure 2. Order of inverse distance weighted averaging: parameter 'm'.

2.3. Differential Settlement Analysis

Differential settlement is calculated to statistically analyze the distribution of the difference in settlement of points on a grid. This analysis shows the roughness and steepness of the slopes in the final surface, which is a critical input to surface water management of the reclamation cover.

Once the surface profile of the consolidated deposit is numerically modeled as indicated in Equation (3), the differential settlement can be calculated and analyzed through different measures that are available in the literature. Several formulations for surface roughness calculations are listed and reviewed by [44]. The same concepts can be implemented to measure the surface roughness. The differential settlement metrics that are used in this paper are listed in Table 1 and illustrated schematically in Figure 3.

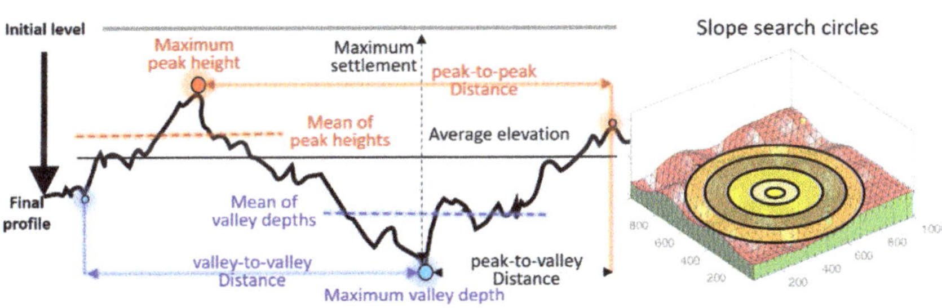

Figure 3. Differential settlement metrics.

Table 1. Differential settlement metrics.

- Maximum valley depth
- Average depth of N valleys
- Average distance between valleys
- Maximum peak height
- Average height of N peaks
- Average distance between peaks
- Average elevation of the surface
- Standard deviation of points' elevation
- Average of absolute differences from the mean
- Maximum settlement
- Average distance between peaks and valleys
- Maximum slope withing search radius

Another simple measure of differential settlement is the 'Delta Z' that merely calculates the difference in the height of any given point and its four adjacent points at its left, right, top and bottom. The distance, in the XY plane, between the given point and the four adjacent points is constant and equal to the grid size. Hence, the distribution and the statistics of the Delta Z is very dependent on the grid resolution. The user can assign the grid resolution based on their site-specific understanding of anticipated deposit profiles.

2.4. Model Validation

The performance of a proposed model can be validated through comparing the model results against the real field data, or the results of available software packages that are already validated through different benchmarks. The 1D model component of the proposed consolidation model is validated by [29,30] through three sets of scenarios, based on the reported cases by [27] as the well-accepted benchmarks. The benchmark cases have been widely used in the literature for evaluation of numerical large strain consolidation models, and are the results of nine different models that predicted the consolidation behavior of four waste clay disposal scenarios [23]. In [29,30], the authors simulated the tailings consolidation under different permeability and compressibility conditions for three types of tailings, i.e., untreated mature fine tailings (MFT), thickened tailings (TT) and phosphate tailings (PT). The authors considered a wide range of solid contents and initial deposit height to investigate the robustness of the model. The primary performance indicators used for validation include settlement over time, effective stress and void ratio profile. The commercial software FSCA is used to compare the key performance indicators for thickened tailings, based on the large strain consolidation theory.

The interface height for the validation cases are illustrated in Figure 4. The 1D model component provides acceptable settlement predictions, labeled TMSim and TMSim-consol in Figure 4, when benchmarked against commercial software and laboratory datasets. Readers are directed to [29,30] for comprehensive details about the validation of the 1D model.

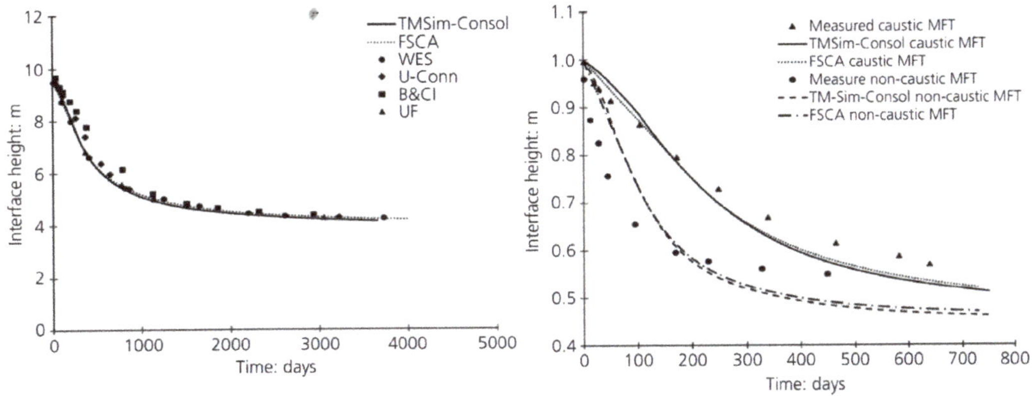

Figure 4. Comparison of settlement and interface height with time for the 1D model—adapted with permission from [30].

The quasi-3D model is tested for special conditions to verify its numerical results. As expected, it generates zero differential settlement (a flat surface) when all of the defined points follow a unique settlement curve. However, there are no available data about the actual surface of a deposit during the active life of a tailings storage facility or post deposition after the pond is filled. A review of the publicly available tailings reports published by oil sands operators clearly states that during the active life cycle of a tailings storage facilities, because the deposits are constantly filled with the tailings, the operators do not track the consolidation: "Settlement and consolidation of fluid tailings are not tracked in active ponds, due to the complexity and inaccuracy inherent in measuring and calculating these values." [45]. Therefore, there are no valid data available to be compared against the results of the model. On the other hand, the developed model here generates a pseudo-3D estimation of the potential surface of the consolidated pond once the active deposition is over and the treated tailings has been consolidated. The model is not intended to predict the final surface profiles or for stage-volume predictions, but rather to inform the closure landform design process.

In the literature, the 2D and 3D models are rarely validated independently from 1D models [46]. Two of the reasons from the literature are: (1) the 2D effects become significant when the impoundment is relatively deep (with a width-to-height ratio in the order of 5 or less, associating with side drainages), whereas the majority of the tailings storage facilities in the oil sands region have much higher width-to-height ratios [47], and (2) while a 1D model satisfies most of the requirements [39], there are limited detailed published data to compare 3D models. Therefore, the developed model utilizes the previously validated 1D consolidation model to infer probable surface roughness or differential settlement statistics across a deposit under different scenarios due to uncertainties.

3. The Case Study

The authors reviewed a number of recent fluid tailings management reports from some of the largest Alberta oil sands operators [45,48–52] that are publicly available and published in accordance with Directive 085 [1]. These reports include tables containing annual estimations of volumetric values of different slurry components such as fines, sand and water that will be generated in oil sands processing and pumped into the tailings storage facilities. The estimated values will be updated on an annual basis once the actual oil sands production and processing happens. The reports also include satellite maps, cross-sectional and plan views of the tailings storage facilities with the sampling point coordinates marked on the maps, all in PDF format. The spatial data, including the sampling data such as the solid contents, specific gravity and the parameters required for consolidation modeling at marked coordinates, cannot be retrieved from the published reports and maps. Therefore, in the absence of publicly available data on the characterization of full-scale tailings deposits, a case study was developed as an analogue for model development that represents a potential real tailings deposit.

The hypothetical case includes the essential characteristics of a tailings deposit, such as the deposit size, the grid resolution, the coordinates of the grid points, the initial elevation of the deposit surface and the parameters used in the 1D consolidation model. An example of the hypothesized grid is illustrated in Figure 5. Larger grid cells (lower resolutions) can be considered for more homogeneous deposits. A 1 km^2 deposit size was chosen to reduce model runtime but could easily be scaled to larger deposit sizes. Given the very large lateral extents of the tailings ponds (several km^2) compared to the depth (50–100 m), boundary effects are expected to be minimal and drainage will predominantly be vertical. In this example, there are a total of 625 grid cells and 25 Defined points in a 1 km by 1 km square deposit.

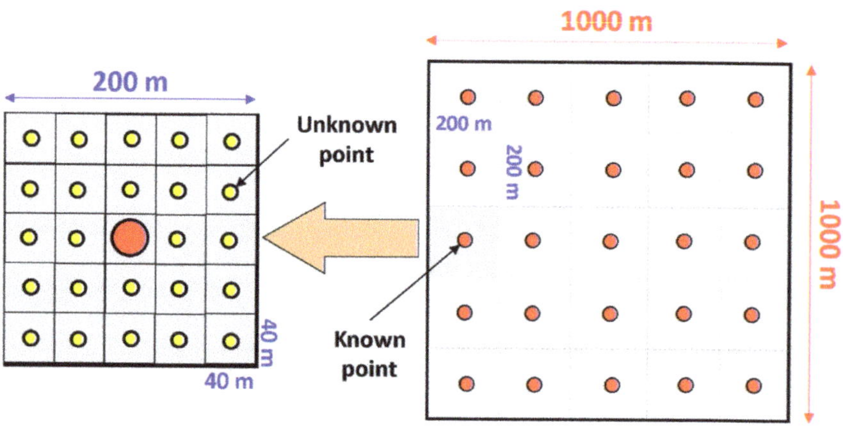

Figure 5. The grid's dimensions and resolutions for the hypothesized case study.

4. Scenarios

The first step of the methodology is to run the 1D model and generate the required settlement curves for the quasi-3D surface estimation runs. The 1D model is run for a total of 67 cases, and the results are investigated to select the scenarios for quasi-3D runs. This resulted in selecting 28 cases for more comparison on the effect of different solids contents and specific gravity (9 cases), surcharge (6 cases) and different material types reflecting a range of A, B, C and D parameters (13 cases). The parameters for these cases are presented in Table 2. Among these cases, A1 and N1 are tailings derived from a centrifuge dewatering process, F1 and F2 are cyclone overflow, C1 and C2 are composite tailings, and the rest are flocculated and thickened tailings from inline or conventional tank thickeners. The centrifuged tailings are representative of tailings with higher fines (~90% fines < 44 μm) and clay contents while the composite tailings are much coarser (~20 to 25% fines < 44 μm) [53].

Table 2. Scenarios considered in the 1D settlement sensitivity analyses.

Investigating	Code	Solids Content	Gs	A (1/kpa)	B	C (m/day)	D
Solids contents (S%) and Specific gravity (Gs)	S1	50%	2.4	1.191	−0.293	2.59×10^{-3}	3.716
	S2	60%	2.4				
	S3	70%	2.4				
	S4	50%	2.28				
	S5	60%	2.28				
	S6	70%	2.28				
	S7	50%	2.65				
	S8	60%	2.65				
	S9	70%	2.65				
Surcharge	T1	60%	2.44	1.620	−0.260	3.50×10^{-4}	3.490
	T2	60%	2.44				
	T3	60%	2.44				
	T4	70%	2.44				
	T5	70%	2.44				
	T6	70%	2.44				
Material types	T7	50%	2.44	1.620	−0.260	3.46×10^{-4}	3.490
	S1	50%	2.4	1.191	−0.290	2.59×10^{-3}	3.716
	A1	60%	2.39	2.220	−0.160	1.73×10^{-5}	4.120
	N1	70%	2.49	2.150	−0.150	1.73×10^{-8}	15.520
	S10	45%	2.65	3.142	−0.290	4.32×10^{-4}	3.686
	S11	44%	2.65	1.311	−0.290	1.73×10^{-3}	3.667
	H1	62%	2.65	0.992	−0.240	3.46×10^{-3}	3.121
	R1	47%	2.65	1.623	−0.260	8.46×10^{-5}	3.398
	G1	47%	2.65	3.140	−0.260	3.46×10^{-5}	3.448
	F1	51%	2.53	2.654	−0.167	3.46×10^{-1}	4.221
	F2	60%	2.53	3.021	−0.293	3.46×10^{-1}	4.000
	C1	50%	2.54	1.090	−0.128	3.50×10^{-3}	4.000
	C2	53%	2.45	1.088	−0.128	3.50×10^{-3}	4.000

For any of these cases, the stochastic 1D model is run for five replications of nested Monte Carlo simulation for 100 years of simulation time. The stochastic A, B, C and D parameters for the inner 1D models, as originally developed by [29], are sampled from BETA_PERT distributions within 10% deviation around the outer A, B, C and D parameters. This deviation envelopes the expected variation of model parameters and is user-adjustable for different cases based on the statistical distribution of the material properties. The inner and outer 'A' parameters are shown in Figure 6. The three groups of settlement curves, each associated with one of the scenarios, are presented in Figures 7–9.

According to Figure 7, as the solids content increases from 50% to 70%, the column of treated tailings settles less over time due to lower void ratios associated with higher solids contents. Figure 7 also shows the settlement is larger for higher specific gravity regardless of the solids content. Figure 8 shows that having a surcharge load will slightly increase the settlement of the column. In both of the two cases (colored in blue and red), the solid lines on top show the situation where there is no surcharge load, the broken lines at the middle represent the cases with two-meter surcharge cap on top of the deposit and the broken lines at the bottom show the cases with a 4 m surcharge cap. However, comparing the two studied scenarios shows that the addition of a 2 m cap does not significantly change the settlement rate over time. That is because there is a very small difference between the settlement of the material with 2 m versus 4 m caps for both solids contents of 60% and 70%. Figure 9 shows how various material types with different constitutive relationship parameters will settle over time. The various range of tailings material in this figure show the applicability of the developed 1D and quasi-3D consolidation models on comparable cases. The scenarios presented in Figure 9 span a large range of tailings material, treated with different technologies and in different stages, including the centrifuge cake, thickened tailings, cyclone overflow and composite tailings. The solids contents and specific gravity of these cases are provided in Table 2. Based on the settlement curves, different materials show one of the five settlement behaviors: the first group includes scenario A1, the centrifuge cake, showing very little settlement over time. The second group of scenarios consists of the thickened tailings, including cases G1 and R1, showing a continuous settlement over time. The third group contains cases F1 and F2, the cyclone overflow, showing a very steep curve in the very early periods, then a flat curve without any more settlements over the rest of the years. The fourth group includes cases C1 and C2, which are composite tailings with a relatively rapid settlement up to year 10, and then continues with a flat curve without any considerable settlement for the rest of years. The fifth group, again thickened tailings material, contains S1 and S11 with a rapid settlement profile with time up to 20 years at which time it plateaus. Figure 10 shows the range in settlement for a particular tailings case with stochastic constitutive relationship parameters that are modeled in GoldSim. The shaded range shows the extent of settlement between the upper and lower bounds, each representing a set of consolidation constitutive relations parameters within the 10% range around the A, B, C and D parameters. The maximum of 5.4 m difference in settlement range in year 40 shows how the settlement is sensitive to the variability of the parameters. Figure 11 presents the constitutive relationship curves to understand the settlement rate of these scenarios as the relations of effective stress, void ratio and hydraulic conductivity for the same scenarios presented in Figure 9. The impact of PSD is captured in these constitutive relationships. Coarser tailings would tend to have higher hydraulic conductivities and lower compressibilities, whereas high fines and clay content tailings would exhibit lower hydraulic conductivities and greater compressibility [53].

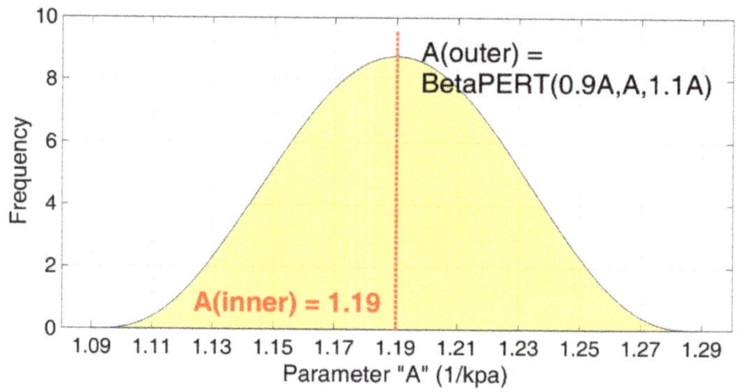

Figure 6. Inner and outer parameter settings for the stochastic 1D model.

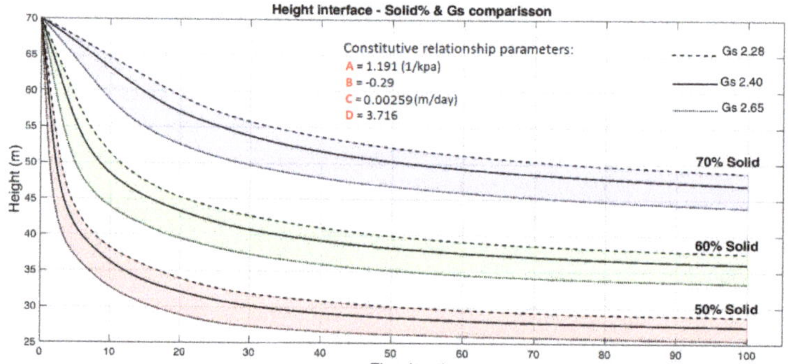

Figure 7. Settlement curves: solids contents and specific gravity scenarios—a thickened tailings case.

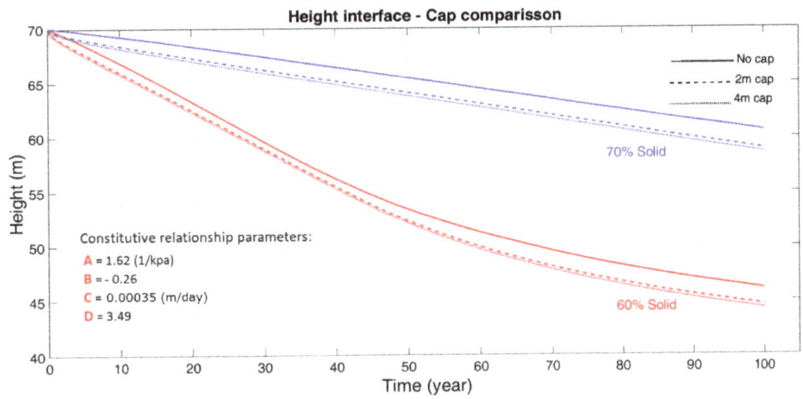

Figure 8. Settlement curves: surcharge scenarios—a thickened tailings case.

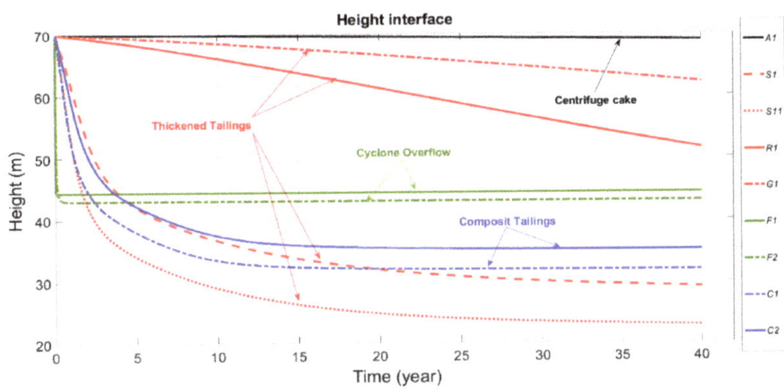

Figure 9. The sensitivity of settlement curves to tailings types.

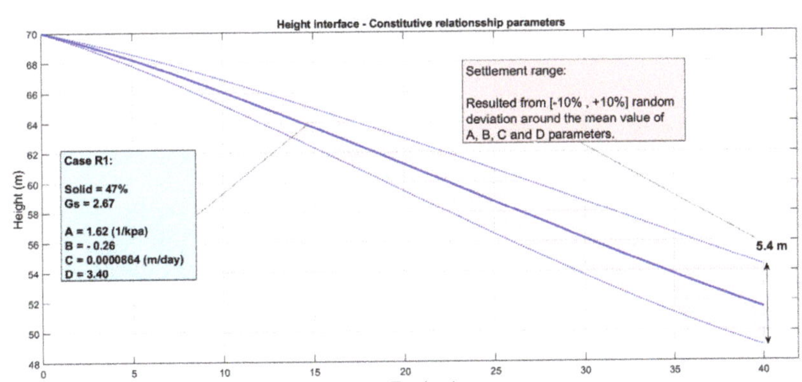

Figure 10. The sensitivity of settlement curves to constitutive relationship parameters.

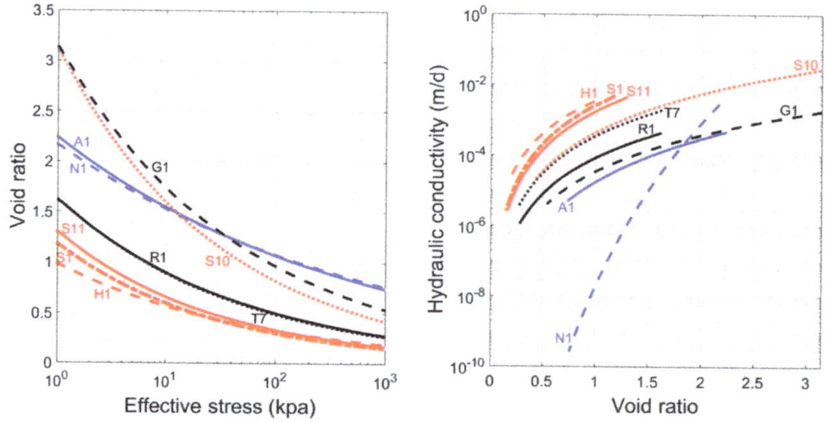

Figure 11. Effective stress, void ratio and hydraulic conductivity for tailings material experiments.

5. Sensitivity Analysis—1D Consolidation Model

The presented scenarios in the previous section shows the sensitivity of the 1D consolidation model and the settlement to some of the selected parameters, such as the surcharge load, constitutive relation parameters, solid contents and the specific gravity. However, there are more parameters involved in 1D and multidimensional consolidation. Reference [54] performed a comprehensive parametric sensitivity analysis to examine if, and to what extent, the consolidation results are sensitive to each of the material property functions and design variables such as compressibility constants, hydraulic conductivity constants, rate of rise, specific gravity, years of deposition, surcharge load application, and initial solids content.

The hydraulic conductivity and the compressibility functions have a combined effect on the consolidation behavior. To capture this combined effect, a full consolidation model must be run that simultaneously accounts for physical characteristics, such as specific gravity and initial solids content, pore water pressure and the stresses associated with tailings deposition. Reference [54] modeled a full consolidation model using the 1D finite strain program FSConsol©. The authors run over 700 models in total, each for a 1000-year simulation time, and the following three results of each model run were recorded for further analysis: (1) height after 1000 years, as an indicator of volume reduction; (2) time to reach 90% consolidation, as an indicator of reclamation timeframes; and (3) solids content after 1000 years, as an indicator of shear strength. Among these three recorded results, the statistical analysis regarding the second measure, i.e., the time to reach 90% consolidation, is presented in Figure 12, after the original work in [54].

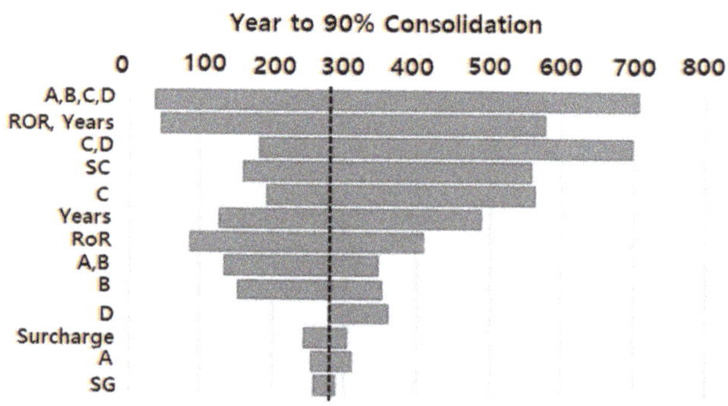

Figure 12. Years to 90% consolidation—after [54].

Figure 12 shows the effect of the model parameters on the range of time to reach 90% consolidation. The first two rows show that changing all of the constitutive relation parameters ends up in a wide range in results. The overall depth of deposit is determined by the Rate of Rise (RoR) and years of depositions (Years) together, and hence, have a significant effect on consolidation times too. The C and D parameters are very influential, and over the studied cases, C has a more significant effect on consolidation than D. The parameters A and B are less influential, to some extent because the 90% consolidation endpoint accounts partially for the compressibility curve. The figure shows that the initial solids content has a relatively strong effect, while the surcharge loading has a very small contribution in consolidation time. The authors conclude that the hydraulic conductivity function is very important in determining the consolidation time, compressibility and hydraulic conductivity inputs have a complex influence on the coefficients of consolidation, and the relationships between rate of raise and the extent of initial consolidation is another important factor in consolidation [54].

6. Quasi-3D Settlement Results

The final step of the methodology in Figure 1 is to generate the quasi-3D surfaces through sampling the settlement curves from the 1D results, generating multiple realizations of the surface profile and analyzing the differential settlement. Among the listed 1D scenarios in Table 2, eleven distinct cases are selected to represent the ranges of different material types, solids contents and specific gravity. The settlement curves of the chosen cases are randomly assigned to the Defined points of the hypothesized grid. The quasi-3D scenarios that are run for 40 years of simulation time and 10 replications are listed in Table 3.

The simulated surface topography is a quasi-3D surface that is generated from replicating the 1D column consolidation model on the 25 defined points, and calculating the simulated Z values at all other undefined points across the deposit surface (1 km by 1 km) based on an extrapolation of at the defined points. The quasi-3D model does not consider the lateral influences among the columns or between columns and facility side walls, as should be considered in real 3D consolidation modeling. This modeling approach has been replicated in the literature, for example in [18,19], and is a valid approach when the goal is to have an early estimation of the potential surface or expected differential settlement in the deposit for long-term decisions where high accuracy of the results is not intended. It is assumed that all 25 columns at the defined points have the same bottom elevation. With this limiting assumption, the simulated quasi-3D surface can be considered as an upper bound for the probable consolidated surface. Additionally, the model assumes the lateral extents are sufficiently large, and hence, the deformation of the side walls of the facility are not considered in consolidation modeling. The real tailings ponds are several kilometers wide with dams 50–100 m high, mainly with one-dimensional drainage. If the pond configurations are different from the assumptions of the presented model, i.e., for ponds with highly variable base topography, relatively small ponds with binding side walls or under valley infill scenarios, the model would not be representative.

Table 3. Summary of the scenarios considered in the quasi-3D analyses.

Investigating:	Case	Solids	Gs
Solids contents	S1	50%	2.4
	S2	60%	2.4
	S3	70%	2.4
Specific gravity	S4	50%	2.28
	S1	50%	2.4
	S7	50%	2.65

Table 3. *Cont.*

Investigating:	Case	Solids	Gs
	A1	60%	2.39
	S11	44%	2.65
	G1	47%	2.65
Material	S1	50%	2.4
	R1	47%	2.65
	F1	51%	2.53
	C1	50%	2.54

Figure 13 shows a simulated surface for scenario G1 at the end of year 40. The grid pixels are 40 m by 40 m.

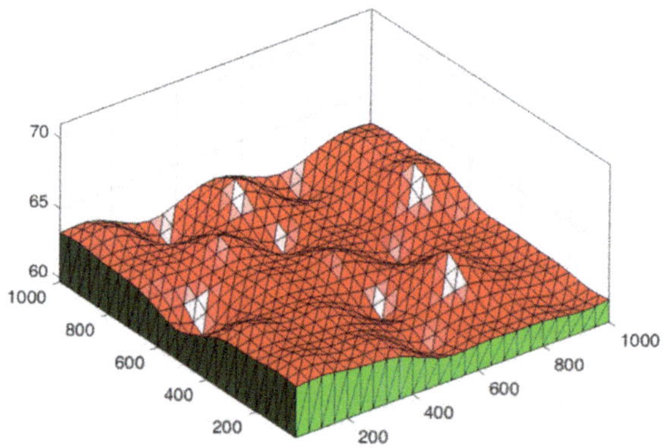

Figure 13. A closer look at the surface and elevation contours: scenario G1, year 40.

The analytical results of the eleven quasi-3D scenarios are compared in Table 4. The reported values show the mean of the metrics over ten replications for each experiment at year 40. Among the listed metrics, the average elevation and the maximum settlement are governed by the 1D consolidation model, while the maximum slope, the standard deviation of elevations around the average height and the roughness show the unevenness of the surface and are more affected by the quasi-3D model. Other metrics, such as the maximum valley depth and peak height and the distances between peaks and valleys, can also be viewed as roughness measures. However, the main metrics that are more representative of the material settlement and surface roughness are the maximum settlement (meters) and the maximum gradient slope (%).

Table 4. Differential settlement analysis for quasi-3D scenarios.

	S1	S2	S3	S4	S7	S11	A1	G1	R1	F1	C1
Maximum valley depth (m)	29.1	39.1	51.4	30.7	26.5	22.7	69.3	60.7	47.7	43.2	31.4
Maximum peak height (m)	29.6	39.6	52.0	30.9	27.0	23.2	69.4	63.9	55.4	46.1	32.7
Mean of 10 valley depths (m)	29.1	39.2	51.4	30.7	26.5	22.8	69.3	61.0	48.3	43.3	31.4
Mean of 10 peak heights (m)	29.6	39.5	52.0	30.9	26.9	23.2	69.4	63.7	54.8	45.7	32.5
Average elevation (m)	29.3	39.3	51.8	30.8	26.7	23.0	69.3	62.3	51.5	44.2	31.8
Maximum settlement (m)	40.9	30.9	18.6	39.3	43.5	47.3	0.7	9.3	22.3	26.8	38.6
Maximum slope (%)	0.4	0.3	0.5	0.2	0.3	0.3	0.1	2.3	5.1	2.1	0.9
St. dev. of elevation (cm)	11.0	8.9	14.4	4.9	9.8	10.6	2.8	72.6	151.4	0.6	0.3
roughness (m)	0.09	0.07	0.12	0.04	0.08	0.09	0.02	0.63	1.19	0.47	0.20
Dist2D_PeakValley (m)	626	597	616	596	648	639	650	633	607	561	564
Dist2D_Peaks (m)	441	457	455	510	428	460	417	464	429	456	456
Dist2D_Valleys (m)	457	451	435	408	466	408	450	440	451	500	491

The results of the three solids contents scenarios are compared in Figure 14A. As expected, the case S1 with 50% solids contents settles the most due to the higher volume of water that it contains and expels over time, compared to the cases S2 and S3. Figure 14B shows the same pattern of maximum settlement values for specific gravity experiments. In this case, the material with the highest Gs (S7) shows the highest settlement because it contains heavier contents (less bitumen), which tends to settle quicker than the other two cases S1 and S4. However, the maximum slope does not show a direct relation to the solids contents or the Gs. The results of tailings material scenarios can help in explaining the maximum slope results.

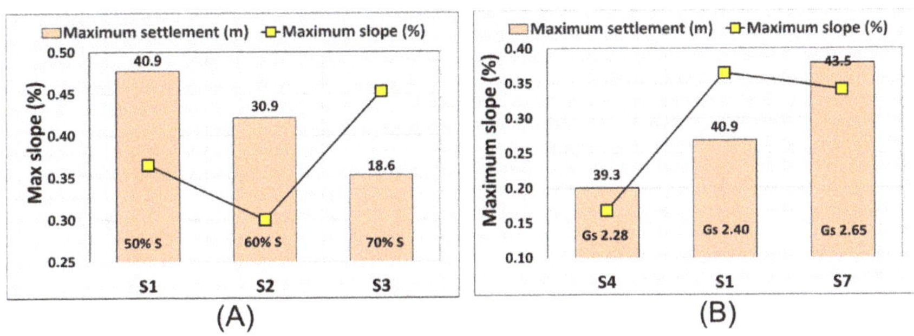

Figure 14. Settlement analysis for different solids contents (**A**) and specific gravity (**B**).

According to Figure 15, the type of tailings material affects both settlement of the thickened tailings over time and the roughness of the final deposit surface. In fact, different constitutive relationship parameters (the A, B, C and D parameters in Equations (1) and (2)) will directly determine the settlement behavior of the material and indirectly the roughness of the final surface of the consolidated deposit.

Among the numerical metrics of the surface differential settlements, slope analysis is an important consideration for closure landform decision making. Assessing the magnitude of slope grades offers insight into the potential future peaks, valleys, surface water flow and possibility for seasonal small and large pools of water. The visual analysis of the quasi-3D scenario results shows that the roughness of the final surface is more affected by the type of tailings materials, rather than the solids contents or the specific gravity. In order to analyze the slope distribution statistically, all the combinations of the points on the final surface are paired, and the slopes of the straight lines between all the pairs are calculated. Then, the slope values are sorted based on the straight distance between the points and grouped in 100-m classes. Figure 16 shows a sample slope distribution as box plots for one of the replications of case R1. The box plots show that the slopes are relatively steeper in short distances (smaller than 400 m). The median of slope values is 1% for a 100-m distance, with a maximum slope of 5.1%, which is also reported in Table 4 and Figure 15. The median of slope values is reduced to around 0.5% for 500 m distances, then to 0.25% in 800 m and remains the same for farther distances up to 1500 m. This means that the large slope values, i.e., the outlier red plus signs around the box plots, are seen in smaller distances around the points. As the distance increases, the steep slopes are replaced with slight slopes. This general pattern is seen in all cases. However, the difference between the more uneven surfaces (such as case R1) and the smoother surfaces (like G1) is the different values of median and the maximum slopes.

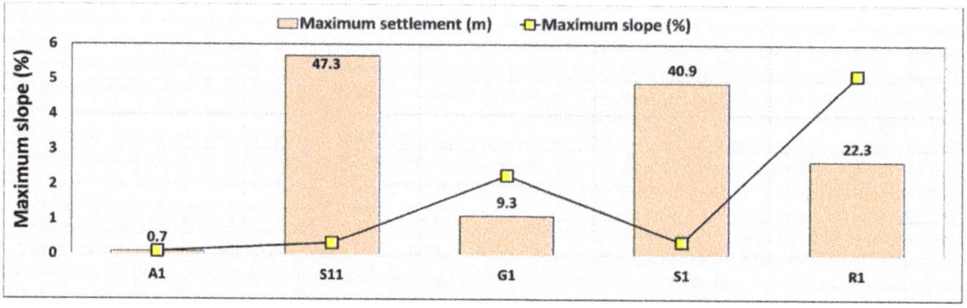

Figure 15. Settlement analysis for different tailings material types.

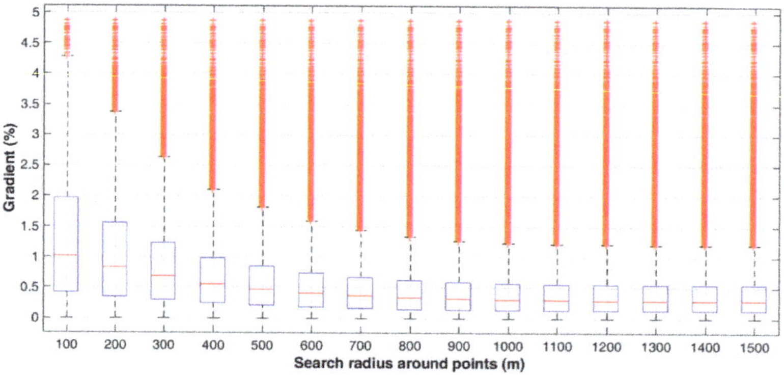

Figure 16. Slope distribution of the final surface: experiment R1, year 40.

7. Conclusions

Integration of closure planning and design through the full life cycle of a mine is a key component of a successful tailings management plan. An early understanding of the range of potential tailings deposit surface roughness that may be encountered is an important input to landscape design for mine closure planning. The main contribution of this paper is the development of a combined 1D and quasi-3D simulation framework to evaluate the surface profile variability of a tailings deposit along with various sensitivity analyses. The 1D and quasi-3D models are developed in GoldSim as a reproducible open-source model that makes them understandable for technical and non-technical stakeholders, rather than typical black-box approaches of commercial software tools. The developed models are implemented under different tailings material type scenarios (thickened tailings, centrifuge cake, composite tailings and cyclone overflow) on a hypothesized grid set.

The numerical results demonstrated tailings that undergo continuous settlement throughout the simulation time (thickened tailings, R1 and G1) will exhibit the largest degree of differential settlement when considering elevation differences between peaks and valleys, greatest maximum slope and potential roughness. Tailings that undergo rapid settlement during time of interest (thickened tailings S1, S11 or composite tailings C1) or very little settlement (centrifuge cake, A1) exhibit much less differential settlement across the deposit. Differential settlements were less sensitive to the tailings properties, such as the specific gravity, which is a function of site geology, and the solids contents for the tailings evaluated (S1).

The developed framework to investigate differential settlement provides insight for landform designers so that designs can be prepared early in the mine life to accommodate the differential settlement or plans developed for ongoing maintenance as needed. These evaluations can provide insight on the impacts to surface drainage features and the potential for unplanned ponding as a result of heterogeneity of the underlying tailings deposit. While the model provides an assessment of the magnitude of differential settlement, it cannot predict the exact locations of flow paths, areas of high and low relief or the gradients of the future surface. Ongoing monitoring and post deposition measurements of the tailings deposit should be conducted and integrated into the closure design process throughout the life of the tailings deposit to update and inform final detailed closure landform design.

Author Contributions: Conceptualization, M.M.B. and N.B.; methodology, M.M.B. and N.B.; software, M.M.B.; validation, M.M.B. and N.B.; formal analysis, M.M.B.; investigation, M.M.B. and N.B.; resources, N.B.; data curation, N.B. and M.M.B.; writing—original draft preparation, M.M.B.; writing—review and editing, N.B.; visualization, M.M.B.; supervision, N.B.; project administration, N.B.; funding acquisition, N.B. All authors have read and agreed to the published version of the manuscript.

Funding: This research was funded by the Natural Sciences and Engineering Research Council of Canada (NSERC) Discovery Grant Program (RGPIN-2019-04573) and Strategic Partnership Grants for Networks Program, "Toward Environmentally Responsible Resource Extraction Network" (NSERC TERRE-NET, NETGP 479708-7c).

Institutional Review Board Statement: Not applicable.

Informed Consent Statement: Not applicable.

Data Availability Statement: Not applicable.

Acknowledgments: The authors used GoldSim Academic version 12.1 from GoldSim Technology Group LLC to develop the Monte Carlo simulation models.

Conflicts of Interest: The authors declare no conflict of interest. The funders had no role in the design of the study; in the collection, analyses or interpretation of data; in the writing of the manuscript, or in the decision to publish the results.

References

1. Alberta Energy Regulator. *Directive 085: Fluid Tailings Management for Oil Sands Mining Projects*; Alberta Energy Regulator: Calgary, AB, Canada, 2017.
2. Alberta Energy Regulator. *State of Fluid Tailings Management for Mineable Oil Sands*; Alberta Energy Regulator: Calgary, AB, Canada, 2020.
3. Government of Alberta. *Environmental Protection and Enhancement Act*; Government of Alberta: Edmonton, AB, Canada, 2000.
4. Johnson, E.A.; Miyanishi, K. Testing the assumptions of chronosequences in succession. *Ecol. Lett.* **2008**, *11*, 419–431. [CrossRef]
5. Gosselin, P.; Hrudey, S.E.; Naeth, M.A.; Plourde, A.; Therrien, R.; Van Der Kraak, G.; Xu, Z. *Environmental and Health Impacts of Canada's Oil Sands Industry*; Royal Society of Canada: Ottawa, ON, USA, 2010; Volume 10.
6. Carey, S.K. Growing season energy and water exchange from an oil sands overburden reclamation soil cover, Fort McMurray, Alberta, Canada. *Hydrol. Process.* **2008**, *22*, 2847–2857. [CrossRef]
7. Meiers, G.P.; Barbour, S.L.; Qualizza, C.V.; Dobchuk, B.S. Evolution of the Hydraulic Conductivity of Reclamation Covers over Sodic/Saline Mining Overburden. *J. Geotech. Geoenvironmental Eng.* **2011**, *137*, 968–976. [CrossRef]
8. McKenna, G.; Mooder, B.; Burton, B.; Jamieson, A. Shear strength and density of oil sands fine tailings for reclamation to a boreal forest landscape. In Proceedings of the Fifth International Oil Sands Tailings Conference (IOSTC), Lake Luise, AB, Canada, 4–7 December 2016.
9. Hyndman, A.; Sawatsky, L.; McKenna, G.; Vandenberg, J. Fluid Fine Tailings Processes: Disposal, Capping, and Closure Alternatives. In Proceedings of the 6th International Oil Sands Tailings Conference (IOSTC), Edmonton, AB, Canada, 9–12 December 2018.
10. Shurniak, R.E.; Barbour, S.L. Modeling of water movement within reclamation covers on oil sands mining overburden piles. In Proceedings of the National Meeting of the American Society of Mining and Reclamation (ASMR), Lexington, KY, USA, 9–13 June 2002.
11. Kelln, C.; Sharma, J.; Hughes, D.; Graham, J. An improved elastic–viscoplastic soil model. *Can. Geotech. J.* **2008**, *45*, 1356–1376. [CrossRef]
12. Leatherdale, J.; Chanasyk, D.S.; Quideau, S. Soil water regimes of reclaimed upland slopes in the oil sands region of Alberta. *Can. J. Soil Sci.* **2012**, *92*, 117–129. [CrossRef]
13. Gibson, R.E.; England, G.L.; Hussey, M.J.L. The Theory of One-Dimensional Consolidation of Saturated Clays I, finite non-linear consolidation of thin homogeneous layers. *Géotechnique* **1967**, *17*, 261–273. [CrossRef]
14. Imai, G. Experimental Studies on Sedimentation Mechanism and Sediment Formation of Clay Materials. *Soils Found.* **1981**, *21*, 7–20. [CrossRef]

15. Znidarcic, D. Predicting the behavior of disposed dredging soils Prediction du comportement des refus de draguage apres. Geotechnical Engineering for Transportation Infrastructure: Theory and Practice, Planning and Design, Construction and Maintenance. In Proceedings of the Twelfth European Conference on Soil Mechanics and Geotechnical Engineering, Amsterdam, The Netherlands, 7–10 June 1999.
16. Jeeravipoolvarn, S.; Scott, J.D.; Chalaturnyk, R.J. 10 m standpipe tests on oil sands tailings: Long-term experimental results and prediction. *Can. Geotech. J.* **2009**, *46*, 875–888. [CrossRef]
17. Qi, S.; Simms, P.; Vanapalli, S. Piecewise-linear formulation of coupled large-strain consolidation and unsaturated flow. I: Model development and implementation. *J. Geotech. Geoenvironmental Eng.* **2017**, *143*, 04017018. [CrossRef]
18. Gjerapic, G.; Znidarcic, D. A Mass-Conservative Numerical Solution for Finite-Strain Consolidation during Continuous Soil Deposition. In Proceedings of the Computer Applications in Geotechnical Engineering, Denver, CO, USA, 18–21 February 2007. [CrossRef]
19. Pinho, M.Q.; Oliveira, W.L.D. Large strain consolidation analyses of fine tailings disposal in mining pits. *REM-Int. Eng. J.* **2020**, *73*, 411–419. [CrossRef]
20. Jeeravipoolvarn, S. Geotechnical Behavior of In-Line Thickened Oil Sands Tailings. Ph.D. Thesis, University of Alberta, Edmonton, AB, Canada, 2010.
21. Coffin, J. A Three-Dimensional Model for Slurry Storage Facilities. Ph.D. Thesis, University of Colorado at Boulder, Boulder, CO, USA, 2010.
22. Fredlund, M.; Donaldson, M.; Chaudhary, K. Pseudo 3-D deposition and large-strain consolidation modeling of tailings deep deposits. In Proceedings of the Tailings and Mine Waste, Vancouver, BC, Canada, 26–28 October 2015.
23. Zhou, H.; Amodio, A.; Boylan, N. Informed mine closure by multi-dimensional modelling of tailings deposition and consolidation. In Proceedings of the 13th International Conference on Mine Closure, Australian Centre for Geomechanics, Perth, Australia, 3–5 September 2019; pp. 287–302. [CrossRef]
24. Geier, D.; Gjerapic, G.; Morrison, K. Determination of consolidation properties, selection of computational methods, and estimation of potential error in Mine Tailings settlement calculations. In Proceedings of the 5th International Conference on Tailings and Mine Waste, Fort Collins, CO, USA, 26–28 January 1998. [CrossRef]
25. Gjerapic, G.; Johnson, J.; Coffin, J.; Znidarcic, D. Determination of Tailings Impoundment Capacity via Finite-Strain Consolidation Models. In Proceedings of the GeoCongress 2008: Characterization, Monitoring, and Modeling of GeoSystems, New Orleans, Louisiana, 9–12 March 2008; pp. 798–805. [CrossRef]
26. Amodio, A.; Zhou, H.; Boylan, N. FLAC Based Modelling of Tailings Deposition and Consolidation. In *Applied Numerical Modeling in Geomechanics*; Itasca International, Inc.: Minneapolis, MN, USA, 2020.
27. Townsend, F.C.; McVay, M.C. SOA: Large Strain Consolidation Predictions. *J. Geotech. Eng.* **1990**, *116*, 222–243. [CrossRef]
28. Engels, J.; Dixon-Hardy, D.; Birch, B. A Realistic Technique to Obtain the Surface Contours of Conventional, Thickened and Paste Tailings Storage Facilities. In Proceedings of the Ninth International Seminar on Paste and Thickened Tailings, Australian Centre for Geomechanics, Perth, Australia, 3–7 April 2006.
29. Zheng, X. Development of a Tailings Simulation Model Using System Dynamics. Master's Thesis, University of Alberta, Edmonton, AB, Canada, 2019.
30. Zheng, T.; Beier, N. Simulation of water storage in a reclamation cover incorporating tailings consolidation. *Environ. Geotech.* **2021**, *40*, 1–12. [CrossRef]
31. Kaminsky, H.; Omotoso, O. Variability in fluid fine tailings. In Proceedings of the International Oil Sands Tailings Conference (IOSTC), Lake Luise, AB, Canada, 4–7 December 2016.
32. Yong, R.N.; Sethi, A.J. Mineral Particle Interaction Control of Tar Sand Sludge Stability. *J. Can. Pet. Technol.* **1978**, *17*, PETSOC-78-04-07. [CrossRef]
33. Mikula, R.; Omotoso, O. Role of Clays in Controlling Tailings Behaviour in Oil Sands Processing. *Clay Sci.* **2006**, *12* (Suppl. 2), 177–182.
34. Well, P.S.; Kaminsky, H. Slurry to soil clay behaviour model-using methylene blue to cross the process/geotechnical engineering divide. In Proceedings of the Tailings and Mine Waste, Vancouver, BC, Canada, 27–28 July 2015.
35. Cerato, A.B. Influence of specific surface area on geotechnical characteristics of fine-grained soils. Master's Thesis, University of Massachusetts, Amherst, MA, USA, 2001.
36. Guo, C.; Wells, P. Some properties of Suncor oil sands tailings. In Proceedings of the International Oil Sands Tailings Conference (IOSTC), Edmonton, AB, Canada, 5–8 December 2010.
37. Wells, S.; Guo, C. Application of block modeling to Suncor tailings characterisation. In Proceedings of the International Oil Sands Tailings Conference (IOSTC), Edmonton, AB, Canada, 7–10 December 2008.
38. Esposito, G.; Nik, R. Sand and fines mixing in fluid tailings at the Shell Muskeg River mine external tailings facility. In Proceedings of the International oil sands and tailings conference (IOSTC), Edmonton, AB, Canada, 2–5 December 2012.
39. Wilson, G.W.; Kabwe, L.K.; Beier, N.A.; Scott, J.D. Effect of various treatments on consolidation of oil sands fluid fine tailings. *Can. Geotech. J.* **2018**, *55*, 1059–1066. [CrossRef]
40. Cargill, K.W. *Consolidation of Soft Layers by Finite Strain Analysis*; United States Army, Office of the Chief of Engineers: Arlington County, VA, USA, 1982.

41. Pollock, G.W. Large strain consolidation of oil sand tailings sludge. Master's Thesis, University of Alberta, Edmonton, AB, Canada, 1988.
42. GoldSim Technonogy. *GoldSim User Manual*; Version 12.1; Issaquah: Washington, DC, USA, 2018.
43. Jeeravipoolvarn, S.; Miller, W.; Scott, D.; Kabwe, L.; Wilson, W. Modeling Effect of Bitumen Extraction Processes on Oil Sands Tailings Ponds. *J. Civ. Eng. Arch.* **2017**, *11*, 48–59. [CrossRef]
44. Tonietto, L.; Gonzaga, L., Jr.; Veronez, M.R.; de Souza Kazmierczak, C.; Arnold, D.C.M.; da Costa, C.A. New Method for Evaluating Surface Roughness Parameters Acquired by Laser Scanning. *Sci. Rep.* **2019**, *9*, 15038. [CrossRef] [PubMed]
45. Canadian Natural. *Horizon Tailings Management Report 2018*; Alberta Energy Regulator: Calgary, AB, Canada, 2019.
46. Ahmed, S.I.; Siddiqua, S. A review on consolidation behavior of tailings. *Int. J. Geotech. Eng.* **2014**, *8*, 102–111. [CrossRef]
47. Bromwell, L.G. Consolidation of mining wastes. In *Sedimentation Consolidation Models—Predictions and Validation*; American Society of Civil Engineers: Reston, VA, USA, 1984.
48. Canadian Natural. *Muskeg River Mine Fluid Tailings Management Report 2018*; Alberta Energy Regulator: Calgary, AB, Canada, 2019.
49. Imperial Oil Resources Limited. *Kearl Oil Sands Mine: Fluid Tailings Management Report for 2018*; Alberta Energy Regulator: Calgary, AB, Canada, 2019.
50. Suncor Energy. *2018 Base Plant Fluid Tailings Management Report*; Alberta Energy Regulator: Calgary, AB, Canada, 2019.
51. Syncrude Canada Ltd. *2018 Aurora North Tailings Management Report*; Syncrude Canada Ltd.: Fort McMurray, AB, Canada, 2019.
52. Syncrude Canada Ltd. *2018 Mildred Lake Tailings Management Report*; Syncrude Canada Ltd.: Fort McMurray, AB, Canada, 2019.
53. Cossey, H.; Batycky, A.; Kaminsky, H.; Ulrich, A. Geochemical Stability of Oil Sands Tailings in Mine Closure Landforms. *Minerals* **2021**, *11*, 830. [CrossRef]
54. Rourke, H.; Hockley, D. Assessing oil sands tailings consolidation parameters relative to long-term reclamation. In Proceedings of the Tailings and Mine Waste, Colorado State University, Keystone, CO, USA, 30 September–2 October 2018.

MDPI
St. Alban-Anlage 66
4052 Basel
Switzerland
Tel. +41 61 683 77 34
Fax +41 61 302 89 18
www.mdpi.com

Applied Sciences Editorial Office
E-mail: applsci@mdpi.com
www.mdpi.com/journal/applsci

www.ingramcontent.com/pod-product-compliance
Lightning Source LLC
Chambersburg PA
CBHW061616210926
51850CB00009B/2340
* 9 7 8 3 0 3 6 5 6 2 7 9 7 *